The Classical Tradition in
West European Farming

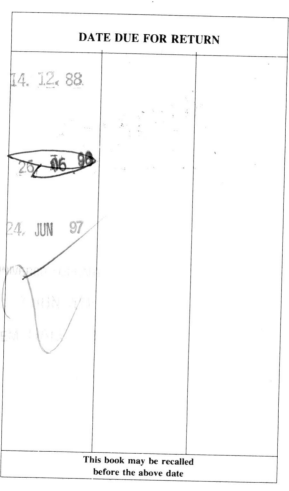

The Classical Tradition in West European Farming

G. E. Fussell *D Litt FR Hist Soc*

President of the British Agricultural History Society, 1968–71

David & Charles

Newton Abbot

ISBN 0 7153 5564 3

Set in 12pt Bembo
and printed in Great Britain
by Latimer Trend & Company Limited Plymouth
for David & Charles (Publishers) Limited
South Devon House Newton Abbot Devon

Contents

List of Illustrations

TEXT FIGURES

Foreword

THERE is a sameness about the processes of nature through the ages, though this has become increasingly disguised with the progress of science and technology; the organic life cycles involved in food production remain the same despite the changes made by the application of scientific chemicals and mechanical equipment. This fundamental truism is a thread that runs through the history of farming processes described in this book.

This study ends intentionally in the mid-nineteenth century, by which time writers of farming works no longer found it necessary to rely upon ancient authority for their precepts. They studied, described and recommended current practice, much of which had been developed by forward-looking men, and was the result of contemporary scientific investigation and mechanical invention. For this reason the classics were no longer studied for practical advice, only for their literary and linguistic value—that is, the *Scriptores rei rusticae*. Farming by the mid-nineteenth century was a modern undertaking, and relied upon its own experience of the conditions in which it was currently working.

My thanks are due to Professor Dr Ernst Klein who kindly read the chapter on the eighteenth century; to the Chicago University Press for permission to use 'Farming Systems in the Classical Era' as the first chapter, this having first appeared in *Technology and Culture*, winter, 1967; to the British Agricultural History Society for the use of that part of Chapter Three that first appeared in the *Agricultural History Review*, February 1969.

Sudbury, Suffolk, 1971   *G. E. Fussell*

CHAPTER ONE

Farming Systems of the Classical Era

Too frequently the debt that modern farming owes to the classical tradition has been obscured and confused in the studies of agricultural historians. The cultivation systems of ancient Greece and Rome were not, to be sure, 'scientific', as some have said. Only by using this term in a sense different from its modern connotation can it have real meaning in relation to the kind of farming practised in prehistoric times in the Middle East and Egypt or in historic times in Greece and Rome. For the ancients had no exact knowledge of the meaning of soil fertility, and they knew nothing of biochemistry in relation to either plants or animals and nothing of the composition of the elements. Their science was a matter of speculation, not of experiment, and could hardly have been applied to what were quite elementary farming processes based upon empiricism. The practical small landed proprietor or peasant worked along traditional lines that were only slowly modified as new ideas of practice occurred to someone and spread by example.

But change there was, pedestrian though it may have been. What little is known of Greek farming from Homeric times to those of Hesiod, Xenophon, and later writers indicates that it was conducted in a rather simpler, less productive manner than that of the Romans, which slowly developed from systems of the same pattern into something slightly more complex and potentially more productive. All that this amounts to is that the Greeks followed a crop and fallow method, whereas, as time went on, some Romans recognised the advantages of a three-course rotation (of a kind), which allowed the land to be cropped for two years in succession and to be fallow the following year or permitted a catch crop of legumes, or the like, to be raised after the grain harvest was gathered and the land to be fallow in the spring. Livestock farming in both countries was almost a form of ranching.

The two contemporary sources for Greek agriculture in the first millenium BC are Hesiod and Xenophon plus a few scattered references in Homer. There were other treatises, but none has been preserved. Both Plato and Aristotle knew of such treatises addressed to the landowners who could carry out their precepts: Theophrastus made use of them. Varro's references to many works now lost have been frequently mentioned. They seem to have been studied by Cassius Dionysius of Utica, who translated the many-volumed work of Mago the Carthaginian (c 550–500 BC) into Greek about 50–60 BC. All later Greek and Roman writers on farming used Mago as a source book. The Roman Senate ordered a Latin translation. Unfortunately the book is lost, allegedly destroyed during the barbarian invasions. The *Geoponika*, too, written in the tenth century AD at Byzantium, was a collection of farming advice made from a number of earlier works, Greek and Roman, many now quite unidentifiable.[1]

The climate and rugged surface of the country were handicaps to the Greek farmer. Crops could be ruined by untowardly heavy rains or hail in the summer or by unusually long periods of sunny weather. Long winters with bitter north winds were another hazard. All farmers have to take their chances with the weather, but the possibilities of disaster in Greece were greater than in a more temperate climate. Elevation, too, was a problem. The high hills, forested or not, could be used for pasturing some species of animals, but the narrow belts of cultivable land in the valleys were liable to damage by spring runoff in times of rapid thaw, especially if the winter snowfall had been heavy.[2]

The Greek farmer's system was necessarily very simple, and he made few changes in the thousand years before Christ. By the time of Homer some tillage for cereal and other crops had already been adopted, though people still remembered the time when cattle were the only wealth. Most of their implements and crops—'the sickle, the wheelless plough, and the cultivation of beans, lentils and onions'—had been learned from Egypt, and from that country or Asia Minor the vine and the olive had been acquired.[3]

As in Central Europe and elsewhere, including America, settled agriculture in Greece depended largely upon forest clearance to open land for cultivation, a process unnecessary in the valley of the Tigris and Euphrates or that of the Nile. Where natural open grassland existed it provided an immediate opportunity for tillage; beyond the limited arable there was some valley pasture for grazing cattle and horses and mountain pasture for breeding sheep and goats. All the modern livestock were being bred—horses, donkeys, mules for transport, oxen for the plough and the cart, cows for milk (though yields were minute), sheep for wool and milk, and pigs, probably the most numerous of all the meat-producing animals. On the arable land barley and wheat were grown, but no rye or oats because the former was a weed of cultivated crops in this area and the latter was unsuited to the climate. The vine and the olive were cultivated widely on a sufficient scale to allow export. Wheat had to be imported to make up for the deficient home supply. Millet was important. Fruit and vegetables were grown. Poultry and many bees were kept.[4]

Hesiod, who lived about a century later than Homer, inherited his father's farm in Boeotia but lost it to his brother in a lawsuit. His poem, 'Works and Days', deals with the farming of his time and place (although it is no more a complete source book than the works of other early writers who followed him at centuries-long intervals). The picture that emerges is of an extremely simple system of mixed farming. Barley and wheat were grown, olives and the vine cultivated. The ground was prepared as well as it could be. There was a chronic shortage of manure. The farmer himself did much of the work and lived well enough in the summer when 'white meats' (dairy produce) were plentiful, but at other times, especially in the weeks just before the harvest, commons were short and famine not unknown.[5]

Already a large part of the common grazing on the mountain pastures was becoming private property and the small landed farmer's lot more difficult for economic reasons—chiefly pressure from the rich. The physical conditions of Boeotia did nothing to alleviate the difficulties. The soil was heavy and intractable, quite opposite to the light soils of Attica. The climate was atrocious in its vagaries. The winter was cold with storms of bitter rain and freezing north winds; snow lay about till the end of spring. Summers were short but intensely hot, with frequent storms. The sole advantage of the climate was that the fertility of the land was partly renewed by the alluvial soil carried down from the mountains.

Hesiod's account of the details of farming is fragmentary: in May, when the Pleiades were rising, the harvest, barley and wheat, was ready. The yield was very small by modern standards, not more than three or four to one, which meant that a man had to plough a good deal of land to secure subsistence for his family. If the ratio was three to one, $2\frac{1}{2}$ bushels of seed would give a yield of $7\frac{1}{2}$, of which $2\frac{1}{2}$ must be saved for seed for the next crop. The remaining five bushels were about half a year's requirements for an adult. A reasonable precaution would be to allow a trifle above the bare needs of the family, as well as something for the livestock. The peasant might set ten bushels per capita as his goal. If married with two children (equal to one adult), he had to sow six acres of land and prepare six acres of fallow for the following year. This acreage plus a few vines and olive trees with pasture for his sheep, goats and pigs on the mountain slopes would be a fairly representative holding.

With a wooden plough not reinforced by iron, preparing the soil meant exhausting work for the peasant. To prepare his seed-bed and fallow he had to go over the land with the plough (which took forty-eight days if he ploughed twice a year, or seventy-two days if he ploughed three times a year), work that was spread over the year in periods of about a fortnight at a time. Harvesting and threshing were other arduous tasks. Cutting six acres of grain with a sickle (made of iron) probably took one man a fortnight. In February and March the vines had to be pruned. In July the grain was winnowed, after being threshed on a smooth floor in an airy place, and was stored in jars. Fodder and litter were gathered to provide for the animals in winter. In September the grapes were cut. When the Pleiades set in November it was time for winter ploughing. A second ploughing was done in the spring, and preferably a third again in summer. This means the fallow got at least two ploughings in preparation for the alternate year's crop. The seed had to be sown while the soil was still well comminuted and not solidified by moisture.

Clearly the peasant with a holding of about fifteen acres could not produce a surplus of any size beyond his own family's requirements for subsistence, and this explains, in part at least, why both Greece and Rome depended largely on imports for their grain supply,[6] particularly wheat; in ancient Greece the production of barley over wheat is estimated at nine to one.[7]

These were the simple practices of the Greek farmer, whether he was a great landowner renting out his land or working it with slave labour, or a small proprietor producing his own subsistence and possibly a minute surplus. Although changes in social organisation were caused by wars, farming techniques were not altered. Land reclamation continued, and forest clearing opened up more fields for cultivation. The arable area was expanded, often for wine and oil rather than breadstuffs. There was good reason for this preference. The vine, the fig and the olive were protectors of the soil. They prevented erosion by the heavy winter rainfall more effectively than the cultivation of cereal crops.

Just when the plough was fitted with an iron share is not precisely known. It was perhaps in the Homeric age, though Hesiod did not mention it; but it was certainly in use when Xenophon was writing only a few centuries later. Xenophon, who lived in the fifth century BC, was a small farmer. He owned a little estate near Scilla. It was isolated but fertile, and Xenophon lived there twenty years, satisfied with farming and hunting.[8]

He was perhaps the first writer to stress the importance of the master's eye. The tenants ought to be watched whether setting trees, tilling, renewing the ground, sowing, or carrying out the fruit. The master should know the nature of his soil and consider the best methods of working it. Fallow ploughing should be done in spring, not in winter because the earth would be slimy, nor in the summer when it was too hard. Ploughing in young weeds then was as good as a dressing of dung. Some Greek farmers in Macedonia and Thessaly ploughed in beans when in flower as green manure. The soil had to be well fallowed to get rid of weeds and open it to the sun. Seed had to be sown at the proper rate, neither too much nor too little, and care taken that it was broadcast evenly. Drainage was by open ditches. When reaping, the crop had to be cut close to the ground if the stalk was short and, if very high, in the middle so that the stubble could be burned to fertilise the ground. Additional dung had to be spread among the ash—weeds rotted in water made good dung. The farmer used horses, mules or oxen for threshing in a carefully chosen winnowing place.[9]

Ploughing in a green crop to increase fertility was excellent advice. Possibly lucern was obtained from Persia and may have reached Greece in 490 BC. The Greeks may have learned of other fodder crops from Babylonia and Egypt, such as clover, beans, peas, vetch and lupines, all of which were indigenous to the Mediterranean area. Cytisus (*Medicago arborea*), a shrubby lucern 'raised extensively in Greece and the Cyclades but little in Italy' were excellent for all animals, especially cows, of which it improved the quality and quantity of the milk yield.[10]

There had as yet been no radical change in farming methods, but there had been some changes in detail. Greek farmers had learned empirically what crops flourished in particular soils, and they also took climate and elevation into account. Their idea of soil varieties was limited to its appearance and physical state, for example, fat, lean, black, sandy, clayey, dry or wet. Similarly, they had preferences in the assessment of the value of animal manures. Dung was mixed with dead leaves and refuse, wood ash for example, to make a compost. Occasional catch crops may have been grown on the fallow, but this practice became more general in Italy and perhaps Gaul at a later date. It is difficult to set a time for any of these innovations. There may have been sporadic changes in cropping, some progress

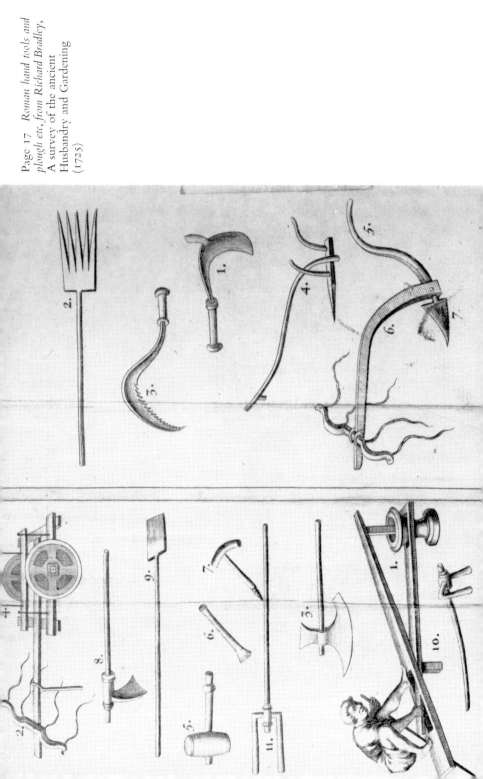

Page 17 Roman hand tools and plough etc, from Richard Bradley, A survey of the ancient Husbandry and Gardening (1725)

Page 18 (above) *Roman ploughshare discovered in London during excavation work in 1964;* (below) *reproduction of the Donnerupland ard made at the Museum of English Rural Life, Reading. This type of plough was used in classical times and later*

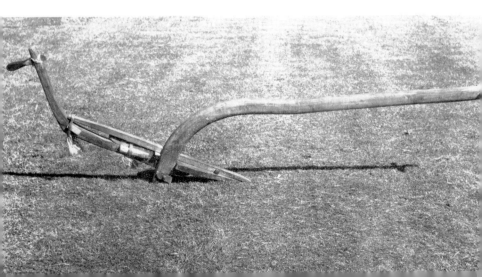

towards a three-course rotation in place of the alternate crop and fallow system, though this change probably took place at a later date and was more marked in Italy. Care must be taken not to exaggerate this change, or the cultivation of the leguminous fodder crops. Many farmers continued in the old ways for centuries.[11]

The Greeks did not demand an iron ploughshare, though some metal shares were being used in the fifth and fourth centuries BC; there were no harrows. There were some large cattle-breeding estates. Dairy farms, small livestock husbandry, and bee-keeping were general, but, despite the statement in many textbooks, actual changes in farming practice were slight. The reason for this is simple: the aristocracy of that time despised the technical arts, and the industry, which was largely the business of slaves, could not therefore bring its aid to the work of the farmer whose idyllic and estimable life was so professedly admired.[12]

When Rome was founded the people were already farmers like the Greeks. They had the same limited range of primitive implements—the wooden plough, the sickle, the axe and other hand tools, but nothing that enabled them to follow any complex system of farming. Their first crops were spelt, barley and

Fig 1   *A primitive type Roman plough used in Spain in recent times*

millet, wheat being introduced later (about 450 BC) according to Varro. Flocks and herds were rare, but all the domestic animals were kept. Holdings were small, but owing to conscription of the peasants into armies the land was neglected and fell into the hands of the aristocracy, who began large-scale farming in the fourth century BC. The population of Rome was then preponderantly peasants whose families worked in the fields to produce their own sustenance. A few owned slaves were assisted by labour remunerated in kind. The growth of large-scale farming followed the Punic Wars, when much was learned from the Carthaginians.[13]

The first Roman writer on farming, Cato the Censor, does not expose this influence in his work, although he has been called a master farmer versed in the lore of the Carthaginians and Greeks, a man who chose crops to cultivate more widely than the working owner of a small farm. The operations carried out on a Roman farm were similar to those of the Jews and Greeks (how in the circumstances could they be otherwise?), but the unwarranted conclusion that the Romans borrowed their technical knowledge from them must not be drawn. Roman agriculture had reached its height in the time of Cato before there was much intercourse between the Greeks and Romans, and when Greek writings had scarcely begun to be studied. It was between the times of Cato and Varro that the Greek and Punic authorities exercised their influence on later Latin authors.[14]

In the second century BC Cato puts more emphasis on the vine and the olive, two commercial crops that were superseding the cereals as market produce in the Campania, parts of Latium, and Samnium. Livestock husbandry took pride of place further south. This development was perhaps the consequence of an increase in population, perhaps of the import of grain from the south—Sicily, Sardinia, Egypt and North Africa.

Cato looked primarily for profit from sales of his cash crops, then to subsistence crops for his slaves, work animals and sheep. His vineyard and olive grove provided marketable produce, but only the net product of grain went away from the estate. In two places Cato set out use of the land of his ideal farm of 100 jugera (62.5ac). The farm should include all sorts of soils, a rather large demand for such a small area. The order of precedence of crops was, first, a vineyard; second, a watered garden; third, an osier bed; fourth, an olive grove; fifth, a meadow; sixth, grain land; seventh, a wood lot; eighth, an *arbustum* ('plantation'); and ninth, a mast grove. The second notation is almost a repetition of this with simple changes of order and more explicit notes on how the crops should be grown and on what soil.[15]

Some thirty chapters of Cato's book are devoted to desultory remarks on the year's work beginning with the vintage. Most of these deal with the vine, the olive, the fig, and fodder crops, much less with the cultivation of the cereals. As soon as the vintage was over, fodder crops were sown: *ocinum* (a mixed crop of ten modii of field beans, two each of vetch and bitter vetch, and possibly some Greek oats), vetch, fenugreek, beans, and bitter vetch. A second and third sowing of these crops followed to give a succession of supplies, and after that the land was prepared for other crops, holes being dug in which to plant olives, elms, vines and figs. The fodder crops were carefully tended, half the available manure being applied to them. If there were olive trees in this ground, trenches were dug round them and filled with another quarter of the manure plus soil. The last of the manure was spread on the meadow in the dark of the moon when the west wind was blowing.

Cato was more concerned with the choice of soil and site for the different crops than with the preparation of the ground for the seed, but he was a strong advocate of ploughing often enough to secure proper comminution of the soil. Autumn and spring ploughings were the normal processes, and the wise farmer ploughed as frequently as he could find time for.

A high, dry, sunny, unsheltered place provided the best soil and most favourable situation for wheat. Spelt could be grown in clay, marshy ground, or the poor red soil still known in Italy as *terra rossa*, which also suited lupines or lentils. Barley was sown in new ground (an equivocal description which may only mean land that had been left fallow for one or more seasons) or in land that would bear a crop every year. Three-month (*trimestre*) wheat could be sown in spring if the autumn seed time had been missed by some accident. Cato advised that turnips, rape and radishes ought to be grown on well-manured or heavy land.

Brehaut believed that the turnips and rape (not kohlrabi as the Loeb Classical Library has it) were for forage, but Cato did not say so; his mention of radishes, a favourite Roman relish, suggests that these two were garden crops.[16]

Next in importance to frequent ploughing was manure. Pigeon's dung was highly esteemed for meadow, garden and grain. All other kinds were carefully saved—goat, sheep and cattle. The dregs of olive oil with an equal quantity of water sprinkled round the trees caused them to flourish, and there were crops, as the Greeks had known, that fertilised the ground —lupines, field beans, vetches and other legumes. The wise farmer made a compost of straw, lupines, chaff, bean stalks, husks, and leaves of evergreens and oaks. Weeds, grass and sedge, dwarf elders and hemlocks were recommended for cow and sheep bedding in order to add to the compost heap. In bad weather the farmer occupied himself by carrying the manure from the cowsheds and sheep pens to the pile, and by repairing his implements and utensils.

A busy time came in the spring when there was a great deal to do in the vineyards, olive groves and orchards; planting, pruning, grafting and so on. It was a time for draining wet land, too. V-shaped trenches were dug, 3ft wide at the top, 4ft deep and 1ft and a palm wide at the bottom. These were paved with loose stones or, if stones were lacking, willow branches or brushwood bundles. The spoil was put back on top of the filling —though Cato does not specifically say so—but it may not have been, as trenches 3½ft deep and 4ft wide were afterwards dug, so placed that the water ran into the ditches. This is the earliest mention of this kind of hollow draining, if hollow draining it was, but it was a reversal of the later practice which was for the hollow drains to run across the field to open ditches at its sides.

There was apparently no selection of sires or dams for breeding livestock. Mating must have been fairly random as the animals grazed practically in freedom in the woods and open spaces. Cato advised feeding work oxen with leaves collected from the elms grown for the purpose, as well as from poplar, oak and fig. Sheep, too, must have green leaves as long as possible. A wise injunction was to feed the sheep on the land where it was proposed to sow crops. Dry fodder, hay, and such like were saved and stored for winter use. Coupled with this exhortation was another to cut the grass crop at the proper time before the seed was ripe, an axiom modern grassland authorities would certainly underline.

The oak woods were scoured for acorns, a job that followed sowing. A great quantity must have been collected. They were soaked in water and a half modius (about a half peck) given to each working ox daily, but if the animals were not working Cato advised putting them out to grass or giving them a modius of waste from the wine press. Days at grass had to be supplemented by 25lb of hay at night when the animals were stabled (?). If hay was wanting, improbable substitutes were evergreen oak and ivy leaves.

Cato advised that all crop refuse, wheat and barley chaff and vetch and lupine husks, be stored to add to the reserves of feed, but this conflicts with the advice to add all these wastes to the compost heap. A matter of choice, no doubt! There must have been a lot of undergrowth in the grain crops in spite of twice hoeing, because Cato suggested that, when storing straw, that with the most grass in it should be kept under cover and sprinkled with salt for future use in place of hay.

In spring, no doubt to revive the farm animals after the winter hardships, a modius of acorns or wine press refuse or soaked lupines plus 15lb (13.08oz/lb) of hay were fed to them, and as soon as the *ocinum* was ready it was used, being pulled up by hand and used as feed till it died. This was followed by vetch, Italian millet and back to elm foliage. Nothing paid better, said Cato, than to take care of the working oxen, and if they were fed on this scale they were fed generously as to bulk though some of the feed was little better than roughage.[17]

This is only a brief synopsis of the farming methods described by Cato in what Rostovtzeff and others have labelled 'scientific farming'. This is much too proud an appellation to attach to methods that, however meritorious, were essentially empiric. Clearly, too, these methods were confined to a particular area, mainly central and southern Italy, where the effects of the Punic Wars had been to concentrate ownership of large estates worked by peasant tenants and where vast open pastures were devoted to grazing on a ranching scale with the flocks and herds protected by armed slaves, the cowboys and shepherds of their day. In the north, too, in the Po valley, Lombardy, Venetia and the Ligurian Alps, the Celtic population was employed mainly in grazing and pasture farming, producing pigs and sheep; but before they could progress to a more settled agriculture they were conquered by the Romans, who doubtless dispossessed them of the most fertile lands. Changes in the methods employed have been remarked, but, other than the substitution of the vine, olive and fig and vegetables, fruit and flowers for the market in place of grain subsistence-farming, there is little indication of what these changes may have been, although some historians make much of the educated classes' probable acquaintance with Greek and Carthaginian textbooks now lost.[18]

The effect of the Punic Wars in dispossessing the peasant soldiers who were absent from their holdings for so long has been commented upon by many modern writers. But conscription of the peasant farmers was not solely responsible for the change from very intensive spade culture to more extensive cultivation. The physical aspect of Italy is very difficult for the farmer, since much of it is mountain that can only be used for summer grazing. The soils of Italy were varied. The Latin plain was, geologically speaking, of recent date, and the volcanic ash was rich in phosphates and potash to support crops. The growth of forest and scrub, which was cleared by the farmers, had supplied humus. By the sixth century BC the Romans were cultivating the area intensively. Since each man's allotment was two jugera (1¼ac), this intense cultivation is readily comprehended. With spade and mattock the early Roman peasant must have laboured amain if he was to supply food for his family. Only by the use of free grazing was it possible for him to live. Later when his holding, or rather his descendant's, was increased to seven jugera (4·375ac) his lot was easier. Even then it was hardly ploughland. A plough could, of course, have been used in just over four acres, but spade and hoe cultivation would have been more intensive, and most probably, though not necessarily, would have given higher yields and allowed more variety in the cropping.

Tenney Frank may have exaggerated in his estimate that

Latium in the sixth century BC was cultivated with an intensity seldom equalled anywhere, but the peasant farmers were certainly obliged to make the most of their small allotments of land. In the practical absence of manure when livestock were few and roamed at free range, this must have led to soil exhaustion. Again, the deforestation of the Volscan and Sabine hills led to erosion; the Campania was flooded by the uncurbed run-off and became a marsh. Some attempt was evidently made to keep this land in cultivable condition, for there are vestiges of drainage works that were not successful, and were neglected. From one-time arable these marshes became poor, wet grazing land supporting a few cattle for meat to be eaten in the cities, sheep for wool, and goats for milk and cheese. There was a system of summer and winter transhumance.

Frank also suggests that with the erosion of the surface soil on the steeper slopes the roots of wheat and barley could gain no hold, while the grape vine and the olive could thrive on this emaciated soil, even in the tufa and volcanic ash left behind. It is true enough that these plants flourish in almost sterile soil when the rock or tufa is broken up and pockets of soil exist. There is little to support these conclusions, but in the light of modern science and world experience they are very reasonable. Soil erosion in Greece at a slightly earlier date had caused emigration to parts of the western Mediterranean. Possibly the increase of population and the essential need for wider areas of cultivation in those days of primitive apparatus and low yields led to the wars of conquest and the building of the vast Roman Empire.[19]

It would be incorrect to say that Italian agriculture had been stabilised by or before the date Cato was writing, but the general outline he sketched remained correct for some centuries. The system was described during the latter part of the century before the Christian era by Varro and touched upon in Virgil's *Georgics* and the pastoral verse of Horace. Columella in the first century AD wrote possibly the best treatise, and Pliny the Elder produced his *Historia naturalis*, a compilation indebted to its predecessors, and perhaps to a vivid imagination, but used as authoritative for many centuries. Palladius, a century or more later, wrote a textbook in the form of a calendar, but it contains little that is original, and, in addition to the earlier treatises, owes something to Gargalius Martialis.

In the first century BC Varro patriotically lauded his country as the best cultivated, most productive and plentifully supplied with natural resources in the world. Rostovtzeff applauded his claims:

Model farms on Hellenistic lines with dense slave populations, cultivating vineyards, olive groves, gardens, fields and meadows then flourished all over south and central Italy. On the pastures of Apulia, Samnium and parts of Latium thousands of sheep, goats, oxen and cows grazed under the watchful eyes of armed and mounted slaves. Villages and scattered farms were to be found in Etruria, Umbria, Picenum and the Po Valley. The tenants of large proprietors produced grain for their own subsistence and neighbouring markets. On the authority of Horace, whose Sabine farm was worked by a steward and eight slaves plus five coloni, Tibullus and Propertius, the *latifundia* of Apulia, Calabria, Etruria, Sardinia and Africa were worked by innumerable slaves and countless oxen and ploughs employed.[20]

Technically there was, and could be, little difference in the methods employed in cultivating crops. With the simple plough, *aratrum*, hauled by two or more oxen, only a limited area could be prepared for sowing. The ownership of the land had no effect upon this, for even if an owner had hundreds of jugera under crops, the land still had to be ploughed in small pieces. Pliny (XVIII. x; vii) reckoned forty jugera of easy and thirty of difficult land to be a year's work for one pair of oxen. Consequently all large farms had to be split up into sections that one plough could deal with, which Varro more optimistically than Pliny, set at one yoke per 100 jugera (I. xix), not physically separated from each other but nevertheless dealt with as units making up the whole. Several ploughs could work at the same time on sections of a large field. Even with modern horse ploughs this has been done. I saw at least half a dozen horse-drawn iron ploughs at work together in the same large field in Oxfordshire in the early years of the present century.

The didactic writers, Varro and Columella, discuss the variety of soils, the crops proper to each, considerations of elevation and slope, preparation of the fallow land (land that had borne no crop for a year after the harvest) and newly broken grassland. Three ploughings remained standard practice for centuries, but sensible farmers occasionally found it advantageous to plough more frequently. There was a good deal of emphasis on the necessity for cross-ploughing and ploughing aslope because the light Roman plough did not, in fact, turn a furrow.

There was always a shortage of manure, and all kinds of animal excreta, including human faeces, were used. Bird dung, especially pigeon's dung, but not that of water or marsh birds, was highly valued, but the order of precedence in fertilising quality was not agreed. It varied in the different books. Household dirt, leaves and waste vegetation, weeds and so on were all added to the compost heap where the animal manure was stored. An oak post was driven into the midst of the heap to ward off adders who liked to make their nests in the warm fermenting mess.

Immediately after spreading, the manure had to be ploughed in and covered to prevent loss by evaporation, so only that number of heaps should be spread each day that could be ploughed in on that day. Columella advised that for autumn sowing the land must be manured in September, for spring any time in the autumn. A jugerum of plain could be treated with twelve loads of eighty modii (pecks), six less than for thin land, and hilly land with twenty-four loads. Judgement was necessary because too much manure was as bad as too little. The first burnt up the land; the second left it cold.

When short of manure, an advantageous practice was to follow that of Columella's uncle (Marcus Columella was probably a Roman landowner with an estate near Cadiz in Spain), who chalked or marled gravelly land, and put gravel on chalky, stiff soil, which produced high yields of grain and a plentiful vintage (II. xvi). Varro, when in command of an army in Transalpine Gaul, near the Rhine, found places where neither vine, olive nor fruit would grow unless the farmers fertilised their land with *candida fossicia creta*, a kind of marl (I. vii). The use of marl, according to Pliny, a method of feeding the earth by itself, had been discovered in Britain and Gaul and had not been neglected by the Greeks, who used a white clay on damp, cold soils at Megara. Both Nisard (1844) and 'A Virginia Farmer' (1913) justly remarked that this material was still used in some parts of France when they were writing.[21] Pliny noted that ash was so much preferred to dung by the farmers north of the Po that they burned stable dung to increase their supplies of ash to fertilise selected crops.

Cato described a lime kiln, but he was not familiar with the use of lime as a fertiliser. He praised the leguminous crops for this purpose, particularly lupines. These and other crops were recommended by all the later writers. Columella said that they would serve any farmer who had no manure. Scattered on poor land in September, ploughed and cut either with the plough-share or spade, this was as good as the heaviest carpet of dung.

When the land had been drained, if necessary, ploughed and manured, the seed had to be sown at rates most likely to give the greatest yield. Recommended seeding rates are given in Table I. These recommended rates are not so very different from medieval and modern practice. For some centuries English farmers have sown between two and two and a half bushels of wheat per acre, or approximately eight to ten modii, equal to a jugerum sown with five or six and a quarter modii. Yields were estimated on an optimistic scale. Pliny said that one peck of wheat seed yielded 150 pecks in soil like Byzacium, Africa (repeating Varro), Sicily and Andalusia. A more likely figure is given for some places in Etruria, where ten or fifteen fold was harvested, but there must have been wide variation, as the parable of the sower and the seed indicates (Matthew 13: 3). In the nineteenth century very high yields were sometimes obtained in Carinthia after slashing and burning.

Among the fodder crops lucern was most highly esteemed, both as forage and for restoring soil fertility. Columella declared that once sown it would last ten years and could be cut four, sometimes six, times a year. Lean cattle grew fat on it; a jugerum would keep three horses a whole year. Pliny was even more enthusiastic about lucern than Columella. He estimated it would last thirty years.

Other fodder crops were *ocinum* (a mixed forage), vetch, cytisus (snail or tree clover) and, of course, lupines. Suitable soil had to be chosen for each. There is some divergence in the

Table I

Recommended seeding rates

Crop	Seed per jugerum: modii			
	Columella II. ix, xx	Pliny	Varro I. xliv	Palladius II–IV
Wheat:*				
Fertile	4	–	5	–
Middling	5	–	–	–
Bearded Wheat:				
Fertile	9	–	–	–
Middling	10	–	–	–
Common wheat:				
Fertile	5	5	–	–
Red-bearded wheat:				
Middling	8	–	–	–
Emmer	–	10	–	–
Barley:				
Six row	5	6	6	8
Two row	6	–	–	–
Panic and millet	4 sex	4 sex	–	5 sex
Beans†	4–6	6	4	–
Lentils	1	3	–	1
Linseed	8	–	–	10
Chichling-vetch	3	3	–	–
Red chick peas (Punic)	3	3	–	3
Sesamum	4–6 sex	–	–	–
Turnips	4 sex	–	–	–
Navew	5 sex	–	–	–
Spelt	–	–	10	–
Vetch	–	12	–	3–6
Lupines	–	10	–	–
Fenugreek	–	6	–	6

Note. Jugerum = 0·625 acres; modius = 1·92 gallons; sextarius = 0·96 pint.

  * One-fifth more seed wheat was necessary among trees.

  † If beans were sown after corn, apply 24 loads of dung; otherwise plant on fat, low, moist fallow.

translations of *ocinum*. A very similar word is *ocimum*, usually translated 'basil', but by the Loeb edition as 'clover'. It was mentioned by Cato, and Varro stated that it was derived from the Greek and used to purge cattle (I. xxxi), whereas Pliny said that it was fed to stop scouring, just the opposite (XVIII. xlii). However, one Manlius Sura had, he said, given another meaning to the word: it was a mixed crop of ten modii beans and two each of vetch and chick peas per jugerum, sown in autumn preferably with some Greek oats as well. *Farrago* was translated as 'barley-fodder' in the eighteenth century, but as 'mixed forage' in the Loeb edition. It was no doubt a mixed crop, a sort of dredge grain, of barley, vetch and leguminous plants sown together (Varro, I. xxi).

In addition to the pulses, legumes and meadow hay used as fodder, turnips, which Cato had implied were a garden vegetable, were apparently being grown in the fields when Columella was writing. He described two sorts, *napus* and *rapa*, both of which filled the bellies of rustics and were good for oxen, being especially popular in Gaul for winter feed. August or the beginning of September was the time to sow the seed in well-prepared and heavily dunged land (II. x). Pliny was so enthusiastic that he placed *rapum* in order of utility precedence after grains and beans. It could be sown either in late February or March as well as in autumn and was good for all kinds of animals. It kept if left in the ground or stored. North of the Po it ranked next after vine and grain and would grow almost anywhere.

Since all these crops could not be grown at the same time in the same piece of land, they had to be cultivated in rotation. The simple crop and fallow system was perhaps the earliest kind of farming and was common to all the Mediterranean area. The bare fallow had to be ploughed frequently in order to conserve soil moisture. This was one reason for turning to the vine, the olive and the fig in Greece. The Romans had passed that stage, at least in some places, by the first century AD, though for this and other reasons the didactic writers from Cato on emphasised the profit to be gained from these crops. The Italian farmer, faced with many difficulties of elevation, climate and soil, had to do what these conditions allowed. The rotation he followed was that of his own choice: he was not compelled to do the same as his neighbours, as was necessary in the open-field communities of the Middle Ages. When it suited him he could leave land uncropped, fallowing in preparation for a more favourable season, or he could seize a chance to take a catch crop of legumes, pulses, or roots.

Although many reputable authorities have stated that the Romans had advanced from a crop one year followed by a fallow to a three-course rotation, formerly believed to have been learned from the Germanic peoples north of the Alps (similar to the winter-grain, spring-grain, fallow system practised at a later date in parts of Germany, France and England), this error has now been corrected. Winter-sown wheat, barley and spelt gave a much greater return for the seed than spring-sown varieties because of the dry summer. Grain and pulse, beans, vetches and so forth were planted in the autumn so that the seed might get the benefit of the moisture. Progress beyond the crop and fallow system took the form of a kind of continuous cropping something similar to what is done in a garden. A crop of wheat or barley was sown in the autumn and harvested in May or thereabouts. The land was immediately ploughed and sown with a catch crop of legumes, lupines for example, which would be off by the autumn and thus allow another winter crop to be sowed. Possibly when this had been harvested the land was allowed to lie fallow for a season. There were farmers who found it necessary to let the land lie fallow for more than one year. This system was not in any sense a three-course rotation. It was an erratic course, much more productive than the early crop and fallow system but determined by the farmer's choice insofar as his environment allowed. For example, where irrigation was possible, three crops a year of grain, legumes and vegetables could be produced between rows of vines as in Campania, where Vesuvius had fertilised the soil, but possibly not in Latium or Tuscany.[22]

Varro began his dissertation about farm animals with sheep, which he thought the first animal to have been domesticated by man. Columella dealt first with the ox, which was used for ploughing, but the sheep and the goat, if not more important, were probably more numerous. Both writers, and Virgil as well, emphasised the necessity for choosing breeding rams and ewes carefully and agreed upon the distinctive features to be taken into account.

Rams had to be well covered with wool from the forehead all over the body, including the belly, and have a broad, long tail. The horns had to be curved and sloping towards the muzzle. If the tongue had any black spots the animal was supposed to be rejected because its offspring would be black or parti-coloured. The ewes were covered in April or, Pliny said, between mid-May and July, yeaning about the end of autumn in warm but not hot weather, when the grass was beginning to spring in the mild showers. Ewes had to be at least two years old before service, better three. They were stabled to yean, and kept with the lambs for two or three days, then taken out to graze, brought back in the evening to suckle the lambs, and separated for the night.

After harvest it was good to put the sheep on the grain fields. They became fat on the fallow ears, and by trampling the stubble and dunging they fertilised the soil. They always had to be properly, even generously, fed, both when in the stall and on the grassland, where they had to be protected from prickly woods, briars and thorns to keep the fleece clean. In the stall, which had to be on a level piece of ground preferably facing east rather than south, care had to be taken to keep their bedding fresh and dry by adding brushwood as necessary, once again to protect the wool and avoid the scab. If the sheep grazed away from the farm, sheepfolds were made of hurdles or netting, the winter pastures often being many miles from those used in summer, but connected by well-worn tracks. Transhumance was normal, the winter pastures in Apulia, for instance, being exchanged for mountain grazing about Reate in summer. The sheep had to have access to clear water and be herded in shady places during the noonday heat. The best grazing times were early in the morning when the grass was moist and after sunset when it became sweet again.

Some fine-woolled sheep were jacketed in skin coats to keep the fleece clean. The farmers of Megara, Attica, and Tarentum were apparently specialists in this product, and exceptional care was taken to cleanse the pens and stables when the animals were kept in the house, the stables being paved to prevent urine from collecting. Besides fig leaves, grape skins, straw, and bran in moderate quantities, lucern and snail clover were considered best for producing meat and milk. One man was necessary for each hundred rough-woolled sheep, two for each hundred fine-woolled sheep.

Goats were important. The best were large with thick legs; a short, full neck; flaccid ears; a small head; and thick, long, black hair. The she-goat was the same, valued also for a large udder and high milk yield. Herds, in Columella's opinion, ought not to number more than one hundred, but Varro recommended no more than fifty. Goats were mated in autumn before December, and kept in cotes with a natural rock floor or paved with stone for ease in cleaning. The kids were dropped as spring came on and feed was blooming. Twins were expected, three kids to two matrons being thought disappointing. The feed of these animals was largely tree and bush browse, but also ivy and shrub or snail clover. They, like the sheep, suffered from diseases, some of which were contagious and might carry off a whole herd.

The sheep and goats were kept for the sake of their milk, wool and hair, not particularly for meat, though doubtless mutton and goat's flesh were sometimes eaten. It certainly would not be wasted. But the prime source of meat in the Roman world was the pig. Excessively fat animals were preferred, boars with square rather than round bodies, pendulous bellies and vast buttocks, short legs, and short, upturned snouts. Sows were chosen for their body length, but were otherwise like the boar. Parti-coloured animals were not as desirable as those of a uniform colour.

Sows were ready to breed at a year old, but were better if not covered till twenty months. Service in February or March allowed the sows to drop their litters in the fifth month thereafter, when the summer feed was luxurious and they were in good milk. The boars were segregated for two months before putting them with the sows. Near the cities suckling pigs were sold; elsewhere the piglets were kept for breeding. Sows continued to produce until they were seven years old. It was important that each sow should suckle her own litter only in her own place, otherwise she would waste milk by suckling any pigling. The second litter was apt to be disappointing. Dropped in the winter when feed was short and the sow's milk consequently limited, the little pigs could easily be runts, thin and small. Wet, marshy pastures were thought particularly suitable, but where there were woods, oak woods in particular, they flourished exceedingly. Mast was collected and stored in cisterns or smoky lofts. The cisterns must have been empty or the mast would have rotted before it was used. Varro praised the great size and excellence of the sides of bacon flitches produced by the Gauls. They were produced in numbers in Lombardy, Narbonne, and eastern Spain.

Ten boars were sufficient to serve 100 sows, though some breeders kept a smaller proportion. Herds ranged from 100 to 150. These animals have always been subject to diseases and one of these from Columella's description seems to have been swine fever. Another was lung trouble.

Already different breeds of cattle could be distinguished. This was natural enough in different countries and climates, so that Asiatic cattle differed from those in Gaul or even in Epirus, but it was also true of Italy itself. The cattle in the Campania were slender, white animals but good at work and ploughing there; Umbrians were also white but much bigger. Here there were also red cattle. In Etruria and Latium the oxen were compact and strong, in the Apennine mountains sturdy and able to endure hardship.

It was generally agreed that the age for breeding was between four and ten years old, but at a pinch two-year-old cows might be served. Cows were kept on short commons for two months before being put to the bull, and the males were highly fed at the same time. Two bulls to sixty or seventy cows was considered the proper ratio. June or July was the month chosen for this operation; it ensured that the calves were dropped when there was plenty of fresh food to be had. Mother and offspring were kept apart at night, and as soon as possible the calves were given some solid sustenance and weaned. Columella supplied rather elaborate instructions for feeding throughout the year in accordance with the supplies to hand, green fodder, if possible, meadow hay, chichling-vetch, vetches, chaff or corn straw,

Page 35   (above) *Model of a Roman ploughman and his plough found at Piercebridge, England, about 100 years ago;* (below) *iron reinforcement of a Roman spade found in London during excavation work*

Inches

Page 36 (above) *Four-ox plough with mouldboard and coulter from a fourteenth-century calendar. Note the wheels;* (below) *fifteenth-century painting of the school of Ferrara showing a ploughing scene. Only two very small cattle were used*

waste products from wine making, oak and green laurel leaves as well as those of the ash, elm and fig. He gave careful directions for breaking in oxen (including cows, I imagine) to work at ploughing and haulage. Cattle diseases are not very clearly described, but various herbal remedies were applied.

A vivid picture of a first-century country villa was drawn by Martial: it was the Villa Faustini at Basum. The house itself was surrounded by open spaces, groves of myrtle, etc, almost a bit of landscape gardening or so it sounds, and there was a garden for pleasure, presumably with flowers, though these were few, only roses and violets, and evergreen and other bushes. Quantities of wheat were grown and filled the granaries, a large stock of wine was produced, and numerous amphoras containing old vintages were kept. Fierce bulls and calves were grazed in a deep valley. The farmyard was full of fowl, geese, peacocks, partridges, guinea fowl from Numidia, pheasants, cocks from Rhodes and other places, doves, gluttonous pigs and gentle lambs. The whole was worked by slaves, including young white slaves born on the estate. Bees may have been kept—Virgil and others discussed them—and wild honey may have been collected. Conical cheeses were made. The diet of these fortunate people was vegetables, eggs, poultry, fruit, cheese and new wine plus the bread staples.

Most wealthy Romans professed an admiration for the simple life of their peasant ancestors of the remote past but were careful to surround themselves with luxury and to refrain from farm work. As the first century advanced this became more difficult. The slave market was almost empty, and there was a general depression of agriculture which caused the tenants to fall into arrears with their rents. Pliny the Younger tried to reduce this loss of income by introducing a system of métayage, a produce rent in place of money. No doubt other landowners did the same.

The suggestion was then, as it has always been, that the harsh, often poverty-stricken and sordid realism of the peasant's life of unremitting toil created admirable traits of character. On these tiny holdings every minute item had to be watched over. The animals, if any, were almost pets, their disproportionate value in the ménage causing them to be cared for as well, if not better than the children. The loss of one of them was a major disaster. The progress of the crops was anxiously watched over from day to day for next year's food depended on the yield. Poets like Virgil, Horace and Martial might have felt differently had they been slaves, tenants, or *coloni* instead of owners. For the peasant, life was always harsh and dour. Anxiety constantly furrowed the sunburnt brow of the hardy men whose animal vigour was so useful in the army and in battle.[23]

Large-scale farming cannot be conducted without proper management of the labour force. Great efforts were made to increase the size of estates and to exploit them on capitalistic lines, but as the supply of slaves became scanty a problem was presented. However large an estate might be, in the current state of technology it had to be cultivated in small portions. Consequently a large ratio of men to area was essential, and each man could only produce a meagre net gain after the seed for the following year was saved and the worker's own subsistence insured. The relatively small number of large cattle kept and the grazing of sheep on the hills and pigs in the forests drastically limited the supply of organic manure. These conditions forced the owners of *latifundia*, especially those remote from a populous market, to domestic economy—production for the family and its clients—and to ranching sheep and cattle.

Métayage on the lines adopted by Pliny the Younger was one way of keeping labour on the land and the land itself cultivated. Tenants on these terms were not altogether dissimilar to those on the medieval manor. Not perhaps legally tied to the land, they were more or less so because the landowner had probably supplied a plough and other tools and livestock including the plough oxen. The man himself was free but had no capital, and the arrangements by which he was enabled to earn a living effectively placed him in a dependant position. He was no doubt the grain farmer, and the contributions of many of his sort made the wealth of the landowner. Between the landowner and the *colonus* there was an intermeditate class, *conductores*, who were responsible to the landowner for his share and supervisors of the work of the *colonus* whom the *conductores* probably exploited.

The spread of the villa system in Spain and Gaul was limited.

38

There were large estates, but many villas may have been no more than fair-sized farms. Augustus had settled veterans on the frontier lands, and the soldiers on service themselves were part-time farmers. These people were *coloni* farming on little more than the subsistence scale of their forefathers. They had no labour management problems: they were, so to speak, self-employed. On the villas there was still this problem, and throughout the first century AD the proprietors of land where reversion to peasant husbandry resting on cereal cultivation was possible preferred to let their land to *coloni*. Marcus Aurelius in the second century encouraged this tendency by settling on Roman soil large numbers of barbarians who were known as *inquilini* and were, it has been said, bound to the land before the *coloni* suffered a similar burden.[24]

Many factors played their part in the growth of ever larger estates and the reduction of the free peasant to the status of *colonus*. In the third century AD Diocletian attempted to bring about stability but cannot be said to have been successful. Only a nominal rent was charged to the hereditary tenant of waste land which the occupier undertook to plant with vines and olives (*emphiteusis*), or the owners of estates were obliged to take over and be taxed for a certain amount of uncultivated land (*epibole*). Anyone prosperous found himself saddled with plots of waste land, which expanded his estate and thus his commitments.

Farming in Gaul was devoted to livestock and cereals, but it had some lessons for the Roman conquerors as they had some for the conquered. The soil of Cisalpine Gaul was extremely fertile in spite of the mountains that were of no use to the farmer. Emilia, Lombardy and Venetia still produce fine crops of wheat today, and in classical times Rome relied on this part of Italy for a proportion of its supplies. The swamps through which Hannibal marched to the north of Parma had to some extent been drained in Strabo's day. Rye was grown in Piedmont, where the natural conditions suited it. Millet and panic grass were grown, the Celts here used the former as food, if not the latter. Barley, perhaps the preponderant Roman bread grain, was extensively cultivated, but it is impossible to say what area was devoted to wheat.[25]

The Romans who lived in Gaul were mainly the army stationed there, officials and the like, who formed but a minority of the population. The soldiers were given land on which to grow their subsistence because the problem of feeding the army, plus export to Rome, was always difficult to solve. There was some centuriation, land being laid out in plots 710 metres square containing 200 jugera separated by public roads into lots 710 metres square, but this affected only a fraction of the cultivated area, which was itself but a small part of the total area of the country. This need for provisioning the forces accounts for the large number of villas in Lorraine and for the prosperity of the Moselle valley. Lizerand suggests that the Romans exercised some influence by precept and example. Narbonne became similar to Italy. The most widespread introduction was the vine to a line extending from Nantes to Paris and the valley of the Moselle.

Something of the same kind happened in Spain. Agriculture had been the foremost industry in that country before the Roman conquest (of necessity as in other countries). The vine and the olive, the former certainly, the latter probably, were introduced either by Greek traders or Roman conquerors. Livestock was a staple, and some surplus grain was produced for export. It has been suggested that Roman influence tended to move communities from the hills to the valleys.

When the Romans penetrated the forest region of northern Europe they found a system of landholding and farming strange to them. It was different from their own, as the natural environment was different. In Caesar's time farming was not new there. Neolithic man had raised crops all over the northern lands to the extent that climate and configuration would allow and necessity demanded. The only written evidence is the inadequate and imprecise report by Caesar and the later one by Tacitus, upon which so many critical commentaries have been written: but this massive literature is mainly constructed of bricks made without straw.

Caesar was a general concerned with living off the country. The proper feeding of his troops was his primary concern, and it was in relation to this problem that he made his few enquiries into and observations upon the farming of northern Gaul and the lower Rhineland. The natives of these parts were mainly devoted to the chase and keeping livestock and lived on milk, cheese and meat. In addition, they practised shifting cultivation on a small scale, but no individual owned land or an estate of his own. The area was apportioned by the leading men to tribes and clans every year, and in such places as they chose. After one year's cropping the men who did the work were compelled to pass on to another piece of land. Caesar's advance came to a stop when his Ubii scouts warned him that the Suebi tribesmen had taken refuge in the forests, because he was afraid he would not be able to collect any more grain for the significant reason that the Germans cared nothing for agriculture. Tacitus confirmed this to some degree when he said that land was taken up by a village as a whole in quantity according to the number of culti-vators and distributed according to status, a procedure simplified by the vast extent of unoccupied land. These Germanic tribes changed the arable land yearly, and there was still a large area of uncultivated land surrounding it. The population did not need to exhaust the natural fertility of the soil by intensive use, there being no vineyards, water meadows, or irrigated vegetable gardens. All they wanted was a grain crop. Nevertheless, he thought the land was fertile in cereals and rich in flocks and herds though the animals were small, presumably in compari-son with those with which he was familiar at home. The con-cern was with number rather than with quality. The diet was simple—wild fruit, fresh venison, and curdled milk. The land was largely covered with forests and unhealthy marshes, and the prime wealth of the time, a sufficiency of iron, was lacking. Tacitus confirmed Caesar's observation that the natives did not care for the monotony of arable farming, preferring warfare, and that they had rather an excessive individuality which de-manded space between houses of simple construction. One opinion is that Tacitus did not know the country but depended on Caesar. Another German scholar suggested that Caesar and Tacitus only knew Germans who had wandered and come into contact with Rome. Neither knew anything of Germans in their homeland.[26] Certainly Caesar only knew them as peoples to be overcome.

Many interpretations have been given to the passages quoted, but the problem remains unsolved and may be insoluble. Field crops and cultivation were not new developments when Caesar's legions reached northern Europe. As Weber observed, all Indo-European peoples farmed, including the German tribes, but he was uncertain, and that uncertainty remains, whether Caesar or Tacitus meant that the Germans moved to a new place every year or merely ploughed fresh land in the neighbourhood every year. His opinion was that Caesar meant the first and Tacitus implied the two-field system. Was this system only wild field grass husbandry at the beginning of our era, or had the three-field system already been invented? As Professor Wilhelm Abel has recently pointed out once again, Tacitus is not clear. Various meanings can be read into his words, which might mean *Feldgraswirtschaft*, two-field or three-field systems, or merely change of ploughed land from year to year. Abel quotes the Salic law as proof that there were then hedged fields because it prohibited breaches and allowing cattle to trespass in grain fields, meadows or vineyards. The crops were rye, barley, oats, wheat, spelt of different varieties, and less frequently millet. Beans, peas, lentils and roots were grown, both in gardens and in fields, as well as flax. The fields were still very small.[27]

Theories are many, but there is no more solid basis for them than the slender evidence of Caesar and Tacitus. It would be a lengthy process to examine them all. Some are based upon inductive reasoning, particularly the belief that the Westphalian *Einzelhöfe* agricultural methods of the eighteenth century were still like those of the earliest times, something that is quite unproven.[28] One thing is certain. Each people must have learned something from the other, in relation to both methods of soil preparation and unfamiliar crops. The Romans, for example, did not recognise rye and oats as crops until they advanced to northern Europe; but they introduced the vine to the Moselle and the Rhine. Such exchanges are normal when foreigners take over a country, or even if they only colonise it. Roman settlers and soldiers lived among the Gauls and Germans, possibly only in comparatively small numbers, just as some Germans and other peoples had been invited to come south by the Romans. They did not get very large areas of land to farm: it varied from two to three and a half acres for a family.[29]

There is little direct information about the technique of farming in Cisalpine Gaul, only a general idea of the crops grown and the domestic animals kept. As much or as little can be said for Gaul, Spain and Roman Britain. Pliny mentions a plough fitted with wheels used in Rhaetia. It is significant that after Pliny's somewhat derivative and often highly imaginative writings there was no original textbook. Palladius had little to say that had not been said before. The assumption must therefore be that there was no technical progress during the first four centuries of our era.

There are two things about Roman Britain that have acquired the quality of traditional folk tales. One is that a heavy, wheeled plough hauled by eight oxen was in use, having been brought to the country by the Belgae; the other is that vast quantities of wheat were grown and exported. Both have been blown upon, if not completely discredited. More than a century ago Thomas Wright remarked that little was known of the state of English agriculture under the Romans but that it was celebrated for its fertility.[30] A little more, but not a great deal, is known today. Caesar's statement on the subject is explicit and may perhaps be quoted once more; the Loeb translation is:

'The inland part of Britain is inhabited by tribes declared in their own tradition to be indigenous to the island, the maritime part by tribes that emigrated at an earlier time from Belgium to seek booty by invasion . . . and after the invasion they abode there and began to till the fields. The population is innumerable; the farm buildings are found very close together, being very like those of the Gauls, and there is great store of cattle . . . There is timber of every kind, as in Gaul, save beech and pine. They account it wrong to eat of hare, fowl or goose; but these they keep for pastime and pleasure. The climate is more temperate than in Gaul, the cold seasons more moderate . . . Of the islanders most do not sow corn, but live on milk and flesh and clothe themselves in skins.'

The Romans had no technical improvements to offer the grain-growing peoples of Britain. They possessed no farm implements superior to those already in use; they used only farmyard manure, household wastes and composts, while in both Gaul and Britain marl and chalk were used in a process of soil

43

mixing that added to fertility. As the occupying armies settled down, and their garrison duties became less and less onerous, the soldiers became more like peasants and probably cultivated land in the same way as their comrades on other frontiers, the Rhineland for example. Villas were built on what might be called the universal plan, similar to those in the homeland and to that of Ausonius in the south of France. No doubt there were establishments of various sizes.

The pre-Roman inhabitants of places like Salisbury Plain and Cranborne Chase continued to live there during the Roman occupation. Some of the new Roman villas were on lower ground, like Chedworth in Gloucestershire, but it is highly problematical whether there was any difference in the processes used by all these peoples in growing crops. Archaeology has made some contribution, but mainly to the discovery of new farm sites, the extension of our knowledge of places where farms existed rather than the technique of the farmers who did the work. Coulters and ploughshares have been found at several places, and from that it has been inferred that both the ard (*aratrum*) and the heavy plough, possibly with wheels, were used to cultivate the land for cereals, bread wheat, spelt, oats and barley. Horses, cattle, sheep and pigs were kept. The Romans, like the Celtic or Belgic people, farmed for subsistence, and the former imported oil and wine. The army of occupation was a limited market for foodstuffs, and the new or improved roads plus peace made for the expansion of the cultivated area and increased numbers of livestock.

It is now agreed that the barbarian invasions did not destroy the organisation of life current in the period of Roman domination everywhere in western Europe. Evidences are the placid life at the villa owned by Ausonius near Bordeaux, the continued existence of schools in the fourth century, and the letters of Sidonius Apollinarius, a native of Lyons, about a century later. The invasions did turn the villas, and western European farming from a system of production based on the export of net produce to Rome and possibly turned the occupying armies back to a subsistence economy with the scale of farming reduced to proportions sufficient to supply the estate and its inhabitants only. Some things the estate could not supply, and there must have been some trading to make it possible to obtain them—luxury goods for the wealthy in the main. There was no improvement in farming technique in all these centuries, although the water-mill was invented and applied to grinding corn. But the number of these mills must have been very small.[31] The so-called Gallic reaper was used in some places early in the Christian era, but to what extent and how long it continued to be used is not known.

Some suggestion has been made that there were two types of farming practised during the classical period. The first was the ploughing of the small, rectangular, so-called Celtic fields of the uplands, but it must be remembered that the Romans them-selves cultivated rectangular fields with the same sort of tools as were used by the earlier peoples.[32] If the long strip was in fact ploughed by the Germanic peoples at this time, the contrast in the farming of Roman Britain is an exemplar of the contrast between the farming of the Mediterranean littoral and that of the countries north of the Alps.

This diversity still exists, but it was perhaps in Spain, or in northern Italy, that the cultivation of the forage crops, the arti-ficial grasses like lucern, sainfoin and cultivated clovers, per-sisted in some small measure, as well as the ploughing in of some green manure. Lucern may, as some have said, have been brought to Spain by the Moors. It spread again through southern France to Burgundy by the sixteenth century and before that date had been revived or reintroduced into the rotations of the Low Countries.

It was the classical practice of taking a catch crop of legumes that formed the basis of modern farming. Growing fodder crops led to taking the plough round the farm unless open-field farm-ing was an obstacle but that was overcome by inclosure in some countries, by agreement in others. The process stimulated a cycle of fertility where formerly the cropping had led to soil deterioration if no worse; but the new fodder crops fed more and better animals which made more and better manure, and the soil comminution that followed row-crop cultivation helped as well. The demands of these crops led to improvements in implements too; but the coming of the mineral and chemical fertilisers of the nineteenth century modified these results as did the development of monoculture. Nevertheless the leguminous crops and careful soil comminution certainly laid the foundation of all modern arable farming, however greatly that may have been affected by modern science and modern engineering. It is the slow progress of these developments during the centuries following the fall of the Roman Empire both in the field but more especially in the literature that it is proposed to trace in the following pages.

CHAPTER TWO

From the Invasions
to the Early Middle Ages

ONLY half-a-dozen Latin textbooks for the farmer have survived. With the exception of the works of Hesiod and Xenophon the methods of Greek farming can only be learned by oblique references in the poets and philosophers. The great work of Mago the Carthaginian has perished, both in its original and translations. The practical farmers who lived between the 'fall' of Rome and the revival of learning in the early Middle Ages doubtless farmed in the ways of their ancestors, but could not have consulted books to guide them for they were not literate, and copies did not exist in sufficient numbers supposing the peasantry could have used them.

Many generalisations have been made, which seem reasonable to a modern mind, but are in fact rationalisations made by the human brain for which there seems to be little foundation. One of these was made by a scholar for whose work I have the most profound respect. Discussing the change to smaller tenant or rented farms that took place in the early Middle Ages, she argued that this change was not always advantageous. The great estates, ecclesiastical or lay, could carry out drainage and reclamation on a large scale, and were the pioneers of progress. This is a good argument. 'It was the great landowners', she continued, 'who studied the treatises on agriculture which had come down from classical times, and it was they for whom new works on the same model were drawn up based in part upon Cato, Varro, Columella or Palladius, and in part upon practical experience'.[1] This is a statement that bears the hallmark of sweet reason, and it is undoubtedly a possibility, but its credence depends upon many unstated factors. Where, and in what number, were there copies of the classical treatises is one question? Another is, who, outside the monasteries, the clerical class trained in them, who administered the great church estates, and such people as the renowned ecclesiastics, could read those that were to be had? How had these books survived the barbarian invasions, and the overwhelming ignorance of almost anything except the polemics of the patristic writers that gave the period before Charlemagne the cognomen 'Dark Ages'? It is, of course, these and cognate questions that I shall try to answer here.

The general works that were produced in the sixth and later centuries do not deal with agriculture in a definite way. It is the accepted opinion that the Elder Pliny's *Natural History* forms the basis of what references there were, and that it was the most widely read of such Latin writings at that time.[2] There was certainly a number of copies and of extracts scattered about western Europe through the centuries. Bede knew the work. Alcuin says that there was a copy of the *Natural History* at York. There was an eighth-century MS containing a part, owned by Robert of Cricklade: a ninth-century copy in an unidentified cloister. King John is said to have owned a copy.[3] This is not very likely to be a complete list. However Laistner holds the opinion that the number of medieval scholars who knew Pliny at first hand is small, an opinion that obviously seems likely to be correct.[4] It is perhaps completely significant that none of the agricultural writers is mentioned in the index to M. Roger's work, *L'enseignement des lettres classiques d'Ausone à Alcuin*.

Scholarship had indeed deteriorated if not disappeared in the West, though a few commentators continued to work, more especially in Italy. The early fathers devoted themselves to the study of the Scriptures rather than to the writings of what was even then classical antiquity. Servius in his commentary on Virgil gave some rather dubious information about Mantuan farming, although his work has been judged as outstanding for his day.[5]

It would be tedious and unnecessary to trace the progress of learning in general through the centuries: but some slight reference to this subject is needed. The extensive range of studies devoted to the subject makes this fairly simple. The favoured reading was the Scriptures, and though there developed a dual attitude towards such pagan literature it was in relation to and explanation of the Scriptures that the pagan authors were studied. A wealth of allegory failed to add any illumination in the realm of natural science. The world as it existed was of less importance than the contemplation of eternity. To occupy oneself with the stars, or the earthly flora and fauna was to question the ways of the Lord God, and that was blasphemous, frightfully dangerous, and liable to lead to eternal punishment. In such a climate of opinion little attention was likely to be paid to what we now call the natural sciences.

The early sixth century was distinguished by the writings of Boethius, Cassiodorus and others. Boethius had intended to translate the whole of Aristotle, but only lived long enough to deal with the logical works. He knew Plato, and like his few learned contemporaries, adapted and, one could say, modernised his sources—possibly to little advantage from a twentieth-century point of view. It seems to be agreed that Cassiodorus (AD 490–585) made the greatest individual contribution to the preservation of learning in the West. Not only did he collect a library at Viviers, but there he encouraged the copying of texts in a well appointed *scriptoria* which was an example to St Benedict, who founded Monte Cassino. He also wished to found a school at Rome to teach humane letters. It is noteworthy that he was well informed about gardening. His library, according to Laistner, contained Latin poets including Virgil, Cicero, etc, 'and finally some medical and agricultural textbooks' which are not specifically named,[6] a most unfortunate omission from my point of view. Singer thought him 'the earliest general writer who bears the authentic medieval stamp'. The scientific heritage is, however, much more fully displayed by Bishop Isidore of Seville (560–636) who produced a cyclopaedia of all the sciences in the form of an *Etymology* or explanation of the terms proper to each. (He only owned one Latin book, Cicero, or so Lot stated). For many centuries the cyclopaedia was most widely read. The works of the series of writers, the Spaniard, Isidore; the Englishman, Bede (673–735) and Alcuin (735–804) and the German, Rabanus Maurus (776–856) who borrow successively from each other and all from Pliny, contain between them almost the entire corpus of the natural knowledge of the Dark Ages.[7]

There was a library at Seville, which, in spite of what Lot said, seems to have been at least as comprehensive as that at Viviers. It contained the works of the Elder Pliny, Solinus, Orosius, Servius on Virgil and other Virgilian scholia, Cassiodorus, Hippocrates, Apulieus, etc, Martial, Palladius and the agrimensores on agricultural and kindred subjects.[8]

Isidore shared, to some extent, the prevalent dislike of profane letters, but did not prohibit their study completely: if he had it would not have been possible for him to compile his *Etymology*. His idea of what was permitted beyond reading and considering the Scriptures was made evident by his great work. Such knowledge was little more than a list of the names of things.[9] Examples are:

Field culture is *cinio, aratio, intermisso, inconsio, stipularii, stercoratio, occatio, runcatio*.

Cinio is burning to get rid of useless moisture.

Aratio is done twice, in spring and autumn.

Stercoratio is spreading manure.

Further references to the different sorts of work on the soil, clearing it of weeds, and covering the seed, are made, and so are the names of the cereals, legumes, vines, trees and aromatic herbs, which were presumably cultivated by the processes named. The livestock, too, are completely catalogued.[10]

Bede naturally used Isidore as a source for his book *De rerum natura* as well as Pliny. He wrote many things besides this, and was a great scholar, necessarily relying on his predecessors in the same field—as we all do, initially at least. Aldhelm (b 639) had some acquaintance with pagan classics as well as the fathers, theological works and the Scriptures. At Canterbury in the seventh century there were 'many works, sacred and profane, hitherto unknown in England'. Unfortunately, Laistner did not supply a catalogue, and whether these included the *Scriptores rei rusticae* remains uncertain. Hadrian, who knew Latin and Greek, is said to have brought to England an abridgement of Pliny in AD 668. Benedict Biscop, a much travelled man, also brought books to England, which Bede studied at Wearmouth.[11]

Alcuin (735–804), like Bede, enjoys a great reputation. He was, Roger wrote, 'in effect the heir of the monks who little by little discovered antiquity'. In the York library there were copies of Virgil, Statius and Lucan, Cicero, Pliny, Pompieus and Aristotle, ie, Boethius. At Lorsh, too, there was a Pliny, a Virgil, Lucan, Horace, Cicero, etc. Fulda possessed a Columella in the ninth century, and other classical and patristic literature. St Gall owned Isidore, Bede and Alcuin, Cassiodorus, Virgil and Vegetius on veterinary medicine. Bobbio had Virgil and Pliny and copies of many of the Latin poets, between 600 and 700 books in all, and Murbach in the Vosges more than 600. Alcuin went to the court of Charlemagne to conduct the palace school, and his work undoubtedly helped to preserve classical learning, though Bolgar, like Singer, rather criticises him, and his close contemporary, Rabanus Maurus (776–856), as merely relying on Isidore and Pliny.[12] In none of this is there any particular reference to the Latin agricultural treatises. Pliny's *Natural History* was fairly well known, and Virgil was read by a good many people, whether the *Georgics* or the *Aeneid* it is difficult to determine. Possibly all his works were read with attention, but that is as much as can be said. The ninth-century library of the Abbey of St Amand in Belgium owned Cassiodorus, Isidore, Bede, Alcuin, Virgil, Pliny, amongst a collection of liturgical works, patristic writings, etc.

Walafrid Strabo (808–49), the author of a life of Charlemagne, was a monk of St Gall. He wrote a poem on gardens, *de cultura hortorum*, recently reproduced in facsimile from a Vatican MS by the Hunt Botanical Library (1966), the last of a number of editions since the invention of printing. It is a description of a ninth-century garden, the plants cultivated and the work done. The garden at St Gall was planned according to 'the traditional layout of the Roman *villa rusticus*, the largest space being devoted to apples, pears, plums, medlars, figs, peaches, mulberries, hazels, walnuts and almonds, and a kitchen garden. Flowers and medicinal herbs were also grown'.[13] At Corbie in Picardy the garden was very large, either divided into four, or else four distinct gardens, and ploughs had to be contributed by certain tenants to keep it in order. Other tenants had to send men from April to October to assist the monks in weeding and planting.[14] The gardeners, though they were somewhat more than that, had continued to follow the layout advised by and carried out by the Romans. Whether this was the consequence of the father teaching the son, worker or owner alike, or whether it was the result of reading textbooks is an unanswerable question. Verbal instruction seems to be by far the most likely: it is a hint that cannot be ignored. Charlemagne has been represented as lighting the lamp of a minor renaissance, which burned brightly for a little while, but flickered out during the troubles of the next century. The management of his vast estates was perhaps brought into some sort of order by the renowned *Capitulare de villis*. The document, one of many capitularies written before the end of the ninth century, has been used by innumerable historians. It does not, however, provide any information about the technicalities involved in producing the food, nor any details of the implements, although Dopsch thought it complementary to the advice given by Palladius. A translation of it is supplied by Guntz, which was taken from Karl Gottlob Anton, *Geschichte der teutschen Landwirtschaft* (1799). A mid-nineteenth-century edition is not considered satisfactory by Latouche. Fraas, in my opinion, valued the *Capitulare* too highly when he said that Carl the Great stands with the Greeks, Aristotle and Theophrastus, as *naturforscher*, or the Romans *de re rustica*, Cato, Varro, Columella, Palladius and the *Geoponika*.[15]

As much or as little can be said for the famed *Polyptyque de St Irminion* (ed Guerard 1844) which was a survey of the properties of the Abbey of St Germain des Prés, or of other similar documents. Useful as these were for administrative purposes, the use for which they were intended, they do not disclose the farming of the day, size of holding, crops grown, cattle kept, and the relations between the lord and his tenants are significant. But they are not records of farming technique.[16] Again the Salic, Bavarian and Burgundian laws supply no more than indirect references to the crops grown—through trespass—the cattle kept—through theft, and so on.

What were the systems of farming followed in the ninth

century, and how much did they owe to the classical textbooks? Of the rustic writers other than Pliny or Virgil, Cato, Varro, Columella, Palladius, there is little record. Unrecorded copies may, of course, have existed and been studied, but there is no certainty about this.

The gardens described by Walafrid and those at Corbie seem to have been rather more than gardens in the modern sense of the word. They resembled the layout of a Roman farm in which fruit trees were carefully cultivated as well as the olive and the vine, and the arrangement often allowed cereals to be grown amongst the trees. The vegetable and herb gardens may have been cultivated on a separate and distinct piece of land, but this is obviously only a guess.

Fraas (it is true that he was writing a century ago) claimed that the German nobility supported a progressive agriculture in the Dark Ages, that the writers used were the foremost Roman authors, and that it could be determined from them which crops, vegetables or fruit had been imported from foreign countries.[17] This is a large claim, and is difficult to substantiate.

It has long been accepted that there was a broad general difference between the cultivation of land in Italy, Greece, and the other lands bordering on the Mediterranean, and the work necessary to be done in countries further north. This resulted in two systems of working the land, but the difference seems to have been one of field-layout rather than of cultivation, although the field arrangements seem to have existed, if not side by side, at least contemporaneously far into the north of Europe, and can be seen in air photographs of the surface of England.

Classical farming was a crop and fallow system. Wheat and barley were the main cereals. The seedbed was prepared with a light plough that did not cut a furrow and invert it, but made a shallow rut in the ground. Bits of earth that fell into this rut were cleared by hand, perhaps broken up with a mattock. The plough was taken over the land three times during the fallow year, and manure was ploughed in before the seed was sown. Wheat was sown in the autumn and ploughed in. There was a variety known as three months wheat sowed in spring, when barley was also put in. The latter was often the more important bread corn. Beans and lupines were cultivated, and both were sometimes ploughed in as green manure. Other crops were in the nature of vegetables, things like peas, radishes and turnips. Millet, panic and spelt was also cultivated. Manure was carefully preserved in a pit into which all vegetables and household waste was thrown including plant residues from the farm. Leaves of elm and ash were cut and saved for winter fodder. Both the olive and the vine were prominent commercial crops. Figs and citrus and other fruits were cultivated. Horses were bred for war; cattle for draught; sheep and goats for meat, milk and wool; cows, of course, were also milked; cheese was made. The animals got little attention beyond being watched by the cowherd or shepherd as they grazed on the waste. Transhumance was practised in the mountains.[18]

This must have been the kind of farming done in Italy from the time of Palladius and Servius's commentary on Virgil until the ninth century when Charles the Great held overlordship of so much of western Europe. Some Greek continued to be spoken in southern Italy because of the influx of refugees in the seventh century and later, but from then, although there were some attempts by the few scattered scholars to preserve classical learning, the country produced no scholars comparable to Isidore, Aldhelm or Bede. The widespread illiteracy can have been no more pronounced than it was in Gaul or Britain.[19] This need not have affected the peasant whose seasonal succession of jobs remained the same as they had been when the *Scriptores rei rusticae* were being written. The monks, who were the caretakers of knowledge, and whose *scriptoria* became very active, are also described as labouring in orchard and field, and perhaps reading more freely the profane classics than formerly. It has been said that they found the right methods by reading Virgil's *Georgics*, and the horticultural experts of antiquity, Gargalius, Columella, Emilian. If they did they were criticised by Gregory, who declared that too many church properties in the Campagna carried sterile cows and useless oxen. This was at the end of the sixth century when the Campagna was slowly declining in physical condition and production. Certainly some monasteries produced a large surplus output a few centuries later, which must have involved the cultivation of very large areas of land. Bobbio, for example, could offer for sale 2,100 bushels of corn (?wheat or other cereals) that could not have been grown on less than 210 acres, and almost certainly occupied more. Hay, oil, pigs and large and small cattle in quantity were also sent to market. The annual dues of a tenant of Viterbo were ten bushels of grain, forty measures of wine, twenty loads of hay and two millstones.[20] He may have seeded one acre to grow the grain, probably at least one and a quarter acres, and cut the hay from possibly twenty acres of waste.[21]

Much play has been made with the physical aspect of the land farmed, density, elevation, rainfall, the conditions to which the farmer had to adapt his processes: and the completeness of the right of user enjoyed by the man who actually did the work. Very little is really known with certainty about the techniques employed in the centuries between AD 500 and 1100. The accepted distinction between the Mediterranean farming (which was and continues in some measures to be the same as that described in the *Scriptores rei rusticae*), and that of the country north of the Alps: in France north of the Loire. In this northern area early farmers are understood to have confined their efforts to the loessal grassland, and did not attack the forest to expand their arable until the fifth and sixth centuries AD.[22]

There is a wide divergence of opinion amongst historians about the effects of the incursions of the Franks and other Germanic invaders into northern Europe in the sixth century. Some believe in an almost complete devastation like that once believed to have been wrought by the Anglo-Saxon invaders of Britain.[23] The opposite view is taken by the majority—at least of those I have read—no very safe foundation on which to build a theory, because each may have been repeating his earlier colleagues instead of expressing original conclusions. So far from devastating the countryside, the barbarians settled down beside the Romano-Gauls in a society which was already highly 'barbarised' before the Franks arrived.[24] Soon after the invasion, it was impossible to tell the difference between the villa of the Gallo-Roman, and the *curtis* of a notable.[25] (I do not intend here to discuss the partition of the land between Gaulish proprietors and the invaders.)

In these circumstances it is unlikely that the villa system of farming would be changed fundamentally because the fields under cultivation were already laid out. The *Lex Salica* demonstrates a concern for the proprietary rights in growing crops, orchards and vineyards, the protection of hay fields. The crops were cereals, navets, beans, peas, lentils, the vine. Fruit trees were cared for. This is a far more complex system than that of the ancient Germans, and is clearly that of a Roman villa based on the methods of farming systematised in the Latin textbooks, though they need not necessarily have been to hand, or read if they were.[26] With such people stock breeding was naturally important. The animals were belled and guarded by day as they grazed on the waste and in the forest. Burgundian law indicates the spread of farming into the woods reclaimed by the very ancient practice of burning. Some villas had disappeared, eg in north-eastern France. They had apparently been abandoned before the invasions, and were possibly settled by groups of barbarians who gave them approximately their modern form.[27]

Wars, plagues amongst men and cattle, depredations by bands of chivalry and footpads all made life less than secure, but in spite of all these uncertainties food had to be produced, and the annual round of work was laboriously followed by the peasants. The organisation of the great estates, royal, ecclesiastical and lay followed the Roman pattern. Part of the estate was retained, or at least the product of that part reserved for the owner. The work was done by the peasants, free and unfree, as part of their payment for the use of the land upon which they produced their own subsistence. The peasant may have used an area of fifteen to seventeen hectares.[28] It was a persistent form of organisation, derived from Roman arrangements, and passing finally into the medieval manor, but there were still some independent villages, and some independent free workers who travelled from place to place offering their services at busy seasons[29] as some migratory labourers did in later centuries.

Improvements in farming systems are said to have been one of the ambitions of owners of great estates held either in large blocks or composed of scattered manors: but there is little basis for this assumption. Any improvement there was in Europe north of the Alps and to the line of the *limes* along the Rhine and Danube was derived from the Romans who occupied all this territory. Their farming has been described in outline above.

They introduced new crops, mainly the vine and fruit, possibly some of the leguminous fodder crops, but essentially their farming must have been conducted on the same system as that of the homeland, especially since it would seem that the Carolingian estates had inherited something of the organisation of the Roman villa. This was a crop and fallow system such as that found in a modified format by Schwerz as long after as the early years of the nineteenth century in practice along the Moselle, where the Romans had introduced vines.[30] And there is little reason to suppose that any new and revolutionary implements and tools were used since the introduction of the eight-ox plough has now been relegated to a much later date than formerly believed, the agricultural tools in use being in general only the hoe, harrow and primitive wooden plough.[31]

St Germain des Prés is one of the best known of these estates. It, like St Amand near Valenciennes, was an ecclesiastical estate. Others of which something is known are the four royal estates in northern France; Annapes and Cysoing, both near Lille; Vitry near and Somain east of Douai. The cropping of these estates is described in some detail by Slicher van Bath, and the number of livestock each carried is known.[32] By inference the rotation at Somain appears to have been two-course, 'in which one-third of the land under cultivation was sown with winter corn, and two-thirds with spring corn.' The area fallowed is not known nor is the shape of the fields. The cultivated land was about 229 to 250 hectares, not including the tenants' holdings. Other examples of estates are the domain of the Abbey of St Pierre, north of Ghent, Prum in south Luxemburg, where the demesne was about ninety hectares. There were many villas in Brabant and Flanders, the Ardennes, Luxemburg, but little or nothing is known of the agricultural technique practised. It may have been the three-course rotation, or rotation of fields or parts of fields for a period in not a few places. There was a two-course rotation as well like that at Somain.[33]

The apparently large area of cultivated land dealt with by the seigneurie through officials and the labours of tenants, even with the areas cultivated by the tenants added, have been described as oases of settlement surrounded by deserts. The total cultivated area must have been negligible compared with that

57

of forest, moor and marsh by which each unit was encompassed. Indeed, on the surface of the estate itself there was often a great area of uncultivated land. Without these areas it would have been impossible for the people, lords, tenants of all kinds and others, to have survived. Not only did these lands supply grazing for the domesticated animals, horses, cattle, sheep, goats and pigs, but the *ferae* could be hunted or trapped, fish could be caught in the rivers, wild nuts and fruit could be collected, and so could wild honey.[34] All this is, of course, perfectly well known.

Little direct evidence has been found to determine how widely the classical traditional and textbook farming survived in Gaul. Consequently very different opinions have been advanced. To these I propose to add my own for what it is worth. So far I have pointed out that survival of classical systems is especially likely on the great estates. Boissonade numbers them as few, but what else was there until the far north-west of Gaul is reached, the Germanic settlements scattered through the dense Hercynian forest, the land slowly being reclaimed from the sea on the Frisian coast, and the Anglo-Saxon settlement of England?

The difference between the smallish rectangular fields of the Mediterranean countries and the long strips of the open arable fields of parts of France, Germany, Flanders and England has been enormously discussed, and is explained in three ways: first, climate, elevation and soil distinctions; second, the use of the contrasting patterns of ploughs, and third, systems of land tenure. These are all intimately connected, and, though they can be expressed in isolation, can hardly be separated in reality. The south was handicapped by the semi-tropical climate, the steep hillsides and mountains, and fairly thin, poor soils, although there were some parts of great natural fertility. In the north the soil was deeper, and less easily exhausted, provided it could be properly comminuted, and there are large areas of fairly level country. The climate was wetter, and whereas the southern farmer must direct all his efforts to maintaining adequate soil moisture, his northern confrere had to get rid of unwanted water. The Roman Conquest involved the settlement of Roman soldiers who had to farm to provide at least a part of their food.

They carried the systems and crops of the *Scriptores rei rusticae* with them.[35] The land allocated to these military colonists, and civilian officials, was laid out in the same form as in their homeland, and their ownership was absolute, but the native farmers, with whom perhaps ideas and crops were exchanged, laid out their fields in the intermixed strips with which every student of agrarian history is familiar.[36] This field layout is supposed by some theorists to be of great antiquity,[37] but this now seems very doubtful. Several other theorists have agreed that it made its first appearance in or about the eighth century AD, and this has to some extent been confirmed by recent research, although much of what is written on this topic is highly speculative, and is not supported by a firm body of documentation.[38]

The origin of this layout is not known. By many people it has been attributed to the use of a plough with wheels constructed in a more massive way than the *aratrum*, and therefore requiring a longer team to haul it through the rather heavier soil of the northern lands,[39] but some doubt has been expressed about the validity of this theory, which has been for so long and so generally accepted. A fertile suggestion has been made by Dr Joan Thirsk. It is that land formerly farmed in singly-owned square or rectangular fields was split up amongst the children or co-heirs of a deceased owner, and that this parcellisation was continued as population increased. 'A multitude of German examples can be cited of townships consisting at one stage of large, undivided rectangular fields, which became divided into hundreds of strips in two centuries or less.' She has found this process taking place in England in the sixteenth century, and in Yugoslavia in the nineteenth and twentieth.[40]

Little is as yet known with certainty about the systems of farming pursued by the Roman colonists in Britain, but it would be surprising if they did not bring with them the methods to which they had been accustomed at home as indeed they did in the *limes* on the continent, and as the Pilgrim Fathers did when they went to America some 1,500 years later. The earlier inhabitants of Britain had also practised farming, and a previous set of invaders, the Belgae, have been supposed to have brought with them a heavy wheeled plough of Germanic or Slav origin that was the tool used in their settlements to set out the cultivated land in the long strips of the open fields.[41] The light plough, the *aratrum* which was in use all over Europe continued to be used, or so it is suggested, in the more remote parts of the country where grazing was more important than grain growing. The existence of this heavy wheeled plough at this time is, however, open to doubt. It is also supposed to have been brought to England by the Anglo-Saxon invaders once again. No site of an Anglo-Saxon village can now be examined because all are said to have been continuously occupied during the past millenium and a half. 'No Anglo-Saxon plough has yet been discovered, nor is there any written evidence about it before the Danish invasions. There is no positive evidence that the settlers brought with them either better implements or better agricultural practices than had already been familiar in Britain for some centuries.'[42]

Fig 2 *An Anglo-Saxon plough*

The whole of this controversy can be said to rest upon the interpretation of the terms *aratrum* and *caruca*, or *araire* and *charrue*, and their contemporary use, a subject that has been discussed by Duby. He suggests that these names for the plough were synonymous in some measure though the symmetrical and asymmetrical ploughs existed side by side in the Carolingian era. What the ratio between them was is now impossible to decide. The heavy plough, with or without wheels, must have been a small proportion of the cultivating implements in use: some of the peasants continued to work with hand tools. The equipment of iron tools to be found on the great estates was extremely limited.[43]

The emphasis laid upon the antithesis between the long strips of the open fields and the rectangular fields of an earlier people, scattered here and there over Europe omits some of the methods practised by the ninth-century farmer. The convertible husbandry (wild field grass husbandry) was fairly widespread, especially in the less fertile, more exposed and hilly country. Land was broken up, and crops sown for a few years on a piece of natural grassland. When exhausted and yielding little, the plot was allowed to fall down to grass again. The so-called infield-outfield system also prevailed in the hills and moorlands. It consisted of two parts. One, the infield, often the more low-lying part of the farm, was cropped every year, and given the benefit of what manure there was; the other, the outfield, was mainly rough pasture, but parts of it were cultivated on the convertible system for some years and then allowed to revert again. Neither of these systems of land use was so general as the crop and fallow system of the Mediterranean countries and some other parts, or as the three-course rotation of the north.[44]

Such were the methods of farming, in outline, that are known to have been practised between the fall of Rome and the rise of the Carolingian Empire, when the first of the new textbooks was written. The first was not a product of the western but of the eastern empire. This was when 'The inherited agricultural textbooks were supplied by new works, and by more or less modernised and embellished translations',[45] but this is much too sweeping a statement because the copies of the 'inherited' books were few and far between, and the new works less than half a dozen written over a period of several centuries. The eastern work was the first of a new series of farming textbooks written by authors of different nationalities between this time and the thirteenth century. Few of the classical textbooks, as already said, seem to have been in the monastic seats of learning, or in the libraries, if any, of the grandees. It has been said that the Greek and Latin masters were ignored, and lay covered in dust in the libraries of the cloisters.[46] The precise date when the *Geoponika* was compiled is uncertain. It has been attributed to the sixth and seventh centuries, but it is more likely to have been produced in the tenth.[47] The author is equally uncertain. Cassius Bassos is believed to have compiled this work: Constantine VII Porphyrogenitus (AD 905–59) had some part in the matter if only to encourage its circulation.

Whoever wrote the book, the author relied heavily upon earlier authority, but unfortunately many of the works of the writers he named have vanished, and some are only remotely connected with farming. Xenophon, Cato, Columella, Pliny the Elder, and Palladius are not referred to, and may therefore have been unknown to the compiler.[48] Varro was known to him. Some of the other authors named are also mentioned by one or other of the Romans without any other details: the works of others remain to us. For example, Democritus of Abdera is referred to by Columella (Lib I. i) and so is Tarentius; Aratus of Soto is known. Cassius Dionysius of Utica was the translator of Mago the Carthaginian according to Varro (Lib I), and supplemented the wisdom of Mago by that of the Greek authors Varro named. Diophanes or Diogenes (Varro) of Bithnyia abridged Mago's twenty books to six, but all these have now vanished. Even this shorter version was cut down once again by Asinus Pollio Trallinanus, but this work, too, is lost. The *Geoponika* also states that Didymus the Alexandrian wrote on agriculture in fifteen books. A good many other names are cited, but are not known or recognisable today.

The *Geoponika* must have been transmitted to the Muslim world just as so many more classical Greek writings were. Most likely it had not been written when the Nestorian school at Edessa was closed in AD 481 and the scholars dispersed.[49] The Arabs, too, had crops and fruits unknown to the western world, and their acquaintance with tropical or sub-tropical farming taught them the uses of irrigation. It was in Spain in all the west where they had the most influence, although they carried their methods to North Africa and Sicily. To these countries the Arabs brought carobs, maize (?), rice, lemons, melons, apricots, oranges, sugar cane, cotton and saffron. Before them the Romans had introduced the culture of the olive and the vine to Spain, and encouraged the production of cereals, etc, mainly with the object of supplying Rome. The restoration of agriculture which took place after the Arabs had conquered most of Spain included livestock, cattle, horses, sheep and goats. Their farming is said to have been much like the Roman—as indeed the climate and terrain demanded, but it was they who introduced irrigation by flooding riparian fields and gardens, and digging canals. Waterwheels were also used to raise the water to a suitable level for distribution.[50]

It was farming of this kind that helped to inspire ibn al Awam to write his *Book of Agriculture*, the Arabic title of which was *Kitab al Felahah*, as transliterated by Clement-Mullett. The Arabic writer also used literary sources, the *Geoponika*, and some of the books on which it is based, but a large number of references is made to a work *The Book of Nabathean Agriculture* compiled by one Abu Baker ibn al Wahshiyya (the European spelling varies). This was at one time said to be very ancient, and purported to be based on the writings of an ancient Chaldean, Qutsâmi, or Kuthami. Al Wahshiyya was quite unscrupulous. He was born in Iraq, and lived about AD 900. On the one hand he has been accused of fabricating sources wholesale, and on the other of making every possible attempt to disguise the fact that the book was based on the Greek and Latin writers on farming and science: yet it was full of information preserved by tradition among the toilers of the Mesopotamian plains. Renan says that it was first 'referred to in Europe by St Thomas Aquinas, was first known amongst Christian scholars, thanks to the quotations made from it by Jewish writers of the Middle Ages, particularly by Moses Maimonides in his *More Nevochim*,' but only in a confused and limited way.[51] Unhappily the translation promised by Chwolson a century ago has not made its appearance.

Ibn al Awam relied on some other Arabic authors on farming and allied subjects, but their works are only known by his quotations. Among them are Abu Umar ibn Hajaj of Seville, ibn al Awam's own birthplace, and other Spanish Moors; but all the information they provide seems to be mainly derived from Greek sources, especially the *Geoponika*. 'J. Ruska has shown that a work by Cassius Bassos, which was translated into Arabic, had a tremendous influence on Moslem agriculturists, and so did ibn Washiya's *Nabathean Agriculture . . .*'.[52] Ibn al Awam, in his preface, declared that he had all the Spanish Muslim farming textbooks as well as the classical authors and, of course, *Nabathean Agriculture*. His work was of a widely different character from the last: but here again the names of the authors on whom he relies have been disguised in the course of their transmission from Greek, perhaps through Aramaic, to Arabic, and thence back to a modern language. Not only was there the language difficulty, but each of the languages employed a different alphabet. Clement-Mullett casts some doubt on ibn al Awam's direct knowledge of the Greeks, Aristotle, Dioscorides, Hippocrates and Galen, Democritos the Abderite, Cassius or Cassianus Bassos of the *Geoponika*. Junius is probably Columella. Varro is doubtful. Anatolius is most likely the same as quoted in *Geoponika*, like Apollonius or Apuleius. One Hadji Khalfa Kastos apparently wrote a book, *Kitab al felihah ar roumieh*, the *Book of Greek Agriculture*, which Clement-Mullett says was translated several times, but is unknown to me. A book of agriculture by al Rhazes is cited by ibn Hedjadj. Sidagos of Ispahan may be Isidore of Seville. Possibly Theophrastus, Virgil and Martial disguised under questionable nomenclature are intended to be named, and perhaps Mago the Carthaginian himself.[53]

The plan of this book owes a debt to Columella and to Varro, whose work was somewhat similar, but on a smaller scale. It is a prototype of the long series of encyclopaedic textbooks of country living, farming and sport that is a feature of a later day. The scale is grandiose, and is blown up by a constant stream of quotation and comment, which becomes exceedingly tedious, although ibn al Awam adds some material gained from his own experience.

Whether the book became at all well known outside Moorish Spain is problematical. For one thing knowledge of the language was limited to a very few scholars, and the book was not translated until many centuries later. One agricultural writer at least had some acquaintance with it. He was Petrus de Crescentius, the thirteenth-century Bolognese lawyer, who wrote the *Liber Commodorus Ruralium* in twelve books. A modest man, he signed it as a compilation.[54] Lastri makes the not quite correct statement that a vast space of time passed after Palladius before Crescentius restored the art of agriculture principally from the *Geoponika*.[55] This book is in fact heavily indebted to the Roman writers, Cato, Varro, Columella and Palladius. It has been said that Crescentius did not know Columella, whose work was not as widely read at this time as that of the other Latin authors. Goetz has said that Crescentius was practically an adaptation of Varro, an opinion shared by Madden, but Crescentius quotes Columella about manuring so he must have read some part, if not the whole of Columella, or been indirectly acquainted with his theories. The *Geoponika* provided a sort of secondhand knowledge of the Greeks. Agriculture was his life, said Fillipo Re, and he determined to bring out an up-to-date textbook, so he combined a reading of all the earlier books he could obtain with information collected on his travels all over Italy and even beyond the Alps. In one respect he is to blame. His book achieved wide fame, was translated speedily into Italian, French and German, and was certainly used as a textbook in Italy and France until well into the seventeenth century. No English translation has ever been made.[56] Despite its popularity in more northern climes, the treatise was mainly adapted to conditions in southern Europe. Olive trees, pineapples, date palms, figs and almonds do not flourish in the harsher climate north of the Alps and the Pyrénées, and the lengthy discussion of their culture was not, as Wilhelm Abel has pointed out, of much use in the temperate countries.[57] This palpable drawback did not prevent the book from being widely read, especially when it was printed some two centuries after it was written, and it is a connecting link between the *Scriptores rei rusticae*, their disciples the author of the *Geoponika* and ibn al Awam, and the authors of the Latin and vernacular textbooks on farming that began to appear all over Europe in the sixteenth century.[58]

The treatises written in thirteenth-century England and so ably edited by William Cunningham and Elizabeth Lamond in 1890, are not quite of the same character. The authors, it can hardly be doubted, must have known of some, if not all, of the Latin writers. They do not disclose this in so many words, by citation or quotation, but their theories and practice are often undeniably similar, although they are weightily concerned with the financial profits to be gained from estate farming, postulating crop seeding and yields, stock feeding and production as the means of making money. A reasonable consequence is that the farming of the peasant for family consumption is totally neglected. The books deal with estate farming, possibly through appointed officers, not only for subsistence, but also on a commercial scale. Contemporary technique must be deduced from the nature of the onerous duties of each employee, which include instructions for ploughing (in Walter of Henley) and the discussion of some fertilisers and their use. Maitland placed Walter in too high a register when he proposed to call him the Arthur Young of his day, enlightened as he may have been.[59]

It is unlikely that there was any number of copies of the *Geoponika* or of ibn al Awam in western Europe except that the first of those works may have been in some libraries in northern Italy, and the second perhaps in Sicily, but there is no evidence to support this speculation. Neither is at all likely to have reached north-western Europe.[60] Crescentius must, however, have had access to these books, either at first or second hand. What did they advise the farmer to do? In the main they followed the precepts of the *Scriptores rei rusticae*.

Like the Roman writers both books are concerned with a simple system of soil classification, not only for purposes of selecting a site for the farm and buildings, but also for the quality and type of soil in which various crops flourished best. Tests by sight, taste and smell were recommended, but Columella thought colour no certain guide. The tenacity of the soil was determined by the classical method of digging a pit (Virgil, second *Georgic*) and refilling it. If the spoil overflowed the hole, the soil was thick and would be sluggish and rise in heavy clods after the plough; if not, the soil was loose and friable, and would yield well. The natural ecology of uncleared land was a useful guide. Both books repeat this advice on choosing a soil. In countries such as those on the north the Mediterranean, elevation is nearly as important as the soil, and the Latin authors divide the land into three classes, plain, hill, and mountain, which was acceptable to the later writers. Crescentius (Book II) repeats these admonitions on selecting a soil. The simple nomenclature and systematisation remained in use until the seventeenth century.[61]

Having chosen the site, or if using land that had been in occupation for some time, the seedbed must be prepared. The Latins from Cato on had always advocated repeated ploughings for this purpose, but in general their idea was that three ploughings at intervals during the year was sufficient, the manure being added before the second so that it might be buried and then mingled with the top soil at the third, which was slight, a mere tumbling of the surface. Clods must be broken behind the plough by manual strokes with a hoe or a mallet (clodding beetle). The seed was hand broadcast and ploughed or harrowed in. Since this system continued to be used in the fields for at least 2,000 years, the new textbooks could do no more than repeat the Latins as much later writers continued to do. Besides being the practice on the semi-arid lands of the Mediterranean littoral, three ploughings were commonly performed in the open fields which had become established in northern Europe by the early Middle Ages. An indication of the people and places for whom the books were written is provided by the assertion made in both the *Geoponika* and by ibn al Awam that the Arabs do not cultivate sandy soil, though the second adds that they occasionally sow barley in this kind of ground.

There is no agreement between the classical writers about the relative value of the different sorts of animal manure and human faeces which were the main standby for fertilising the soil. The differences are not particularly serious, being more the order of precedence than anything else. The droppings of geese and aquatic birds were thought to be harmful, on which the nineteenth-century use of Peruvian guano is a striking commentary. The refuse of the dovecot was uniformly praised, but it should be dried and broadcast with the seed as what might be designated an early form of granulated fertiliser. To add to the always scanty supply, it was recommended that a dung hill or dung pit ought to be constructed to which all the available ordure should be added. Here all household waste, sweepings, ash, and night soil must also be deposited. Weeds, haulm, leaves and other vegetable waste not only increased the supply of manure, but added to its value. This reeking mass should be stirred with a pole from time to time so that it might all rot down. It must not be used the first year, but must be kept till at least the second, better still three or four years old, when it has exhaled the disagreeable smell (with valuable chemical elements), and become mellow. Marl was used in Gaul at the beginning of the Christian era, and the farmers may have practised soil mixing in northern

Europe. Marc Bloch thought that there was an interval of centuries in the use (or disuse) of marl in spite of all the other advice being sedulously followed. Crescentius said that clay and marl make good vines.[62]

The Latin writers were moderately enthusiastic about the advantages of ploughing in green manure, lupines or beans in flower, for example. This process was recommended by the *Geoponika*, ibn al Awam and Crescentius. It may have been continued in Italy where it must have been common form in classical times, and done as a consequence of the inertia of traditional methods until the Middle Ages. Crescentius quotes Avicenna as his authority for advising the practice. He recommends the rate of application of animal manure *per jugerum* advised by Columella.[63] Lupines grow as far north as Britain, and one may speculate whether the Roman army of occupation introduced the idea to north-western Europe. If so the irruption of the Anglo-Saxon hordes put an end to it, and it has never been effectively revived, although lupine green manure was valued in the light lands of east Prussia in the eighteenth century. In spite of all this some land in Italy was under crop only two years out of seven, which sounds like the convertible husbandry, while better soils were cropped two years in every three in the thirteenth century.

All farmers choose to sow what they consider the best seed, that likely to yield most heavily. Seed wheat must be full, firm, smooth and golden in colour: barley much the same. Some farmers choose the largest ears to provide themselves with seed: some tested germination by soaking or sowing a small manured plot, but these must have been few. It was wise to get seed from a different soil since the change would stimulate growth. All this remained wisdom through the centuries.

Again the Latin writers stated the kind of soil they considered suitable for the different crops, or the crops that could be sown with a prospect of success on the soil of the farm. Wheat, for example, ought to be cultivated on the best ground, clayey and moist, barley on dry ground, less productive than that used for wheat, a choice that is quite unexceptionable. Peas, beans and other pulse would flourish on dry ground, lentils and naked barley on dry soil, panic in sandy and so on. Millet, black vetches, spelt, rice, sesame, flax and hemp were amongst the other crops. Ibn al Awam advised his readers to sow beans, chick peas, fenugreek, lupines and other crops on irrigated land. The olive and the vine were very important all over southern Europe, and the vine in France and Germany as it still is. Little or nothing is said about crop rotation in the medieval treatises, but the best time of year for planting is given in the calendars appended to the works. It is worth noting that Crescentius recommends sowing oats in February or March on the grounds that Albertus Magnus said they were good for horses, oxen, donkeys and mules, and could be used by man. This is probably mere copying, it being unlikely that oats were grown in thirteenth-century Italy: the crop was cultivated in the neighbourhood of Cologne just as rye was throughout Germany.[64]

Before passing on to harvesting Crescentius should be given the credit of having used Albertus Magnus (1206–80) *de Vegetabilibus* to postulate something about plant nutrition and physiology. The origin of these ideas was the Greek theorists, derived from Aristotle and Theophrastus, possibly through the medium of Nicholas of Damascus.[65] Albertus was a widely travelled man, and some of his observations were made from nature as well as being stimulated by his enormous breadth of reading. Crescentius, too, was an acute observer of nature as he saw it in his travels. The plant, he assumed, was composed of several elements. Consequently its food ought to contain several. It was some moist mixture, but as plants have no stomach their food must be predigested. The soil forms a substitute stomach, and creates corruption and putrefaction (presumably partly derived from the manure) round their roots. This material is drawn into the plant through the roots, and circulates through the parts thus causing growth. He also tried to classify plants into a system by reference to their parts and outward appearance, and he believed that in appropriate conditions one plant could change into another, eg *triticum* into *siligo*.

Amongst other similar encyclopedists Bartholomew Anglicus was an outstanding example, though he owes a debt to Albertus, and somewhat naturally to Pliny via Isidore. He mentioned that part of the straw (stubble) was burnt to amend the land, and followed the usual path about the crops grown and the animals kept, as well as describing the duties of the oxherd.[66]

Corn was cut in three different ways by the Roman farmers according to the Latin writers. The Umbrians cut the straw close to the ground, and let the sheaves lie as cut. When enough sheaves (I am doubtful about this word: the sense may be only what was cut at one stroke or blow) were accumulated, the harvesters went over them, cut off the ears, and threw them into baskets for convenience in carrying them to the threshing floor. The straw was left in the field to be collected and stacked later. This reaping was done with a hook or sickle.

The method at Picenum and other places described by Varro has led to a good deal of speculation. The tool used here was a curved piece of wood with a small iron saw fixed at the end. The reaper who used the tool simply cut off the ears and left the straw to be cut later. No exact idea of what this tool was like has come down to us. The third method was to grasp as many stalks as possible in the left hand, and cut off the ears above. The ears were immediately carried to the threshing floor, and the straw was harvested later. This could have been done with a sickle.[67]

The second implement may have been the *mergas* of Columella. The drag hooks either beaked or toothed mentioned by him were most probably scythes fitted with a cradle. But this is guesswork. Somehow the *duas mergites* of Pliny, and the *mergas* of Columella seem to have become confused with the curved wooden *bacillum* fitted with the little iron saw described by Varro. I incline to the opinion that the tool referred to was similar to the *pik en hak* used in Belgium in the fourteenth century, and clearly delincated in several MSS, notably the Grimani Breviary. Really this is two tools, not one. The cutting instrument is a straight handle with a blade at right angles like an elbow. This was held in the right hand. In the left was another straight handle with a piece of iron (rod?) also fixed to its end at right angles. With it the reaper pulled a number of stalks towards him, and sheared them off with the cutting edge in his right hand. It is possible, but unproven, that Varro and Columella were both referring to this equipment. It was used for many centuries in Belgium and Holland, and even in Latvia, and in the late eighteenth and early nineteenth centuries an un-successful attempt was made to introduce it into England.[68]

The famous Gallic reaper described by Pliny and Palladius is not mentioned by the later writers, and may have disappeared during the numerous invasions of that country. It was not known to Crescentius. The sickle and the scythe were both used in the early Middle Ages: the *pik en hak* in some countries. Meier

Fig 3   *An imaginative reconstruction of the Gallic reaper referred to by Pliny and Palladius*

Helmbrecht, whose father used four oxen in his plough, brought presents from court, amongst them a whetting stone and a scythe for mowing hay, a hatchet and a hoe. This was in a very present hope of coming into the property.[69] The grain was threshed out of the ear with the flail that had become a jointed stick by this time, but in France and other countries treading out the grain was still the practice of some farmers for many centuries.

Livestock management among the Arabs was pretty much a hit and miss affair, except perhaps for horses. Ibn al Awam has a large section devoted to horse diseases and their cure. It de-rives, at least in part, from the *Geoponika*, Vegetius and Colu-mella. Some of the crops were used for fodder. The *Geoponika* recommends that cows should be driven into the lupines before they were in flower; later the plant would have a bitter taste, and the cattle would refuse it. Cut and sprinkled with sea or river water, lupines would become sweet, and could be dried and used as chaff. Lucern was grown for forage and given to the animals. It was sown in early October, and could be cut as often as discretion permitted. With care one seeding would last twenty years. Crescentius also recommends feeding millet to pigs. Meadow hay, best cut on riparian land, or where winter floods increased the growth, was mown in June, and supplemented by collecting leaves for fodder in the manner the classic writings advised.

All the present-day domestic animals were kept with the exception that ibn al Awam does not mention pigs. Horses were very important for travel and war, but were not often used for farm traction, oxen, donkeys or mules being used in the plough. This author put out 'a veritable treatise on hippology'. The later treatises owe much to the Latins. The *Geoponika* advised farmers to keep their cows low before service, and to choose wellmade heifers with a body of due length, proportionate breadth, good horns, wide forehead, black eyes, compact jaws, flat nose with open nostrils, long and strong neck, straight rather thick legs, good breast, blackish lips, deep flanks, and so on. Bulls should not be allowed to serve until they were two years old, nor after they were twelve. Cows should be of the same age. Mid-spring was the time for service, which should take place when a north wind was blowing to get a bull calf, a south wind to get a heifer. There are many other similar superstitions.

One man and a boy could look after 120 sheep. The ram should be of compact make, carrying long wool and good horns, a not very exhaustive or precise description. Sheep were sheared in mid-spring, and folded in summer in a shady place. At other times of year they spent at least the night in a cote of a size to give each animal ample space, and keep them warm and dry on a shelving pavement pitched with stones. Sheep and goats' milk cheese was made.

A pig of a uniform colour was better than a variegated one, a highly dubious idea. One boar was sufficient for ten sows, but service should be arranged so that none farrowed in the winter. At that time of year the sow had little milk, her teats were cold, which the little pigs did not like, and the dams were inclined to drive them away. The piglings were left with the mother for two months, being kept separately after that. These animals were chiefly fed on acorns, but fattened on wheat bran, the refuse from threshing. Barley was also used for fattening, and making the sows fit to breed.

The three textbooks so far dealt with were written for use in southern and eastern Europe, and possibly North Africa. The Arabic possessions in Asia are outside the scope of this study. It is odd that at about the same time as Crescentius in Italy several writers were producing textbooks of estate and farm management in England and perhaps even more odd that no such treatises appeared in the flourishing agricultural and industrial communities of the Netherlands.

The English works do not usually make specific reference to the classical authors, but this is not likely to mean that the authors were not familiar with the ancient works. The English writers are concerned with current practice, and a good deal of what they say about the technical processes of farming is apparently the expression of contemporary knowledge and well-established tradition. Walter of Henley, *Seneschaucie*, the Rules St Roberd, as well as the few chapters of *Fleta* that deal with estate management all begin by advising a careful survey of the lands, buildings, etc, belonging to the estate, plus the manorial rights. They are largely concerned, too, with the qualifications that should be looked for in the various officers, steward, seneschal and so on down to the humble ploughman, shepherd, etc, who actually did the work. It is evident from what they say that very careful supervision was essential to secure good workmanship, something that could not have been a novelty then, or in classic times, and is no more so at the present day.

Walter of Henley made it quite clear that the lands he had in mind were laid out in the open field strip system, which had, by this time, become very general all over Northern Gaul, parts of Germany and Midland England. He estimated that a plough could work eight or nine score acres a year according to whether the lands were arranged in two or three fields. If two, both winter and spring corn were grown in one, the other lying fallow in alternate years: if three, one part was for winter seed, one for spring, and the other lay fallow once in every three years. This pronouncement is repeated by *Fleta*.

By this time there was some idea that using horses for plough-ing would get the work done more quickly, but the ploughmen were against this unprecedented haste. They took care to keep the pace of horses down to that of oxen. They did not want to walk further over the clods than they did behind an ox-drawn plough. *Fleta* agrees. Too much has been made of this supposed change, which began to be initiated, it has been said, in the eleventh century or even earlier.[70] Walter and his contemporaries began an age-long controversy that lasted until the nineteenth century about the comparative merits of ploughing with horses or oxen, but in spite of the manifest advantages of horse traction,

Fig 4   *A twelfth-century plough, showing oxen harnessed by rope*

oxen continued to be used to within a hundred years of our own time, and are in some remote places still. Walter expressed the opinion that a mixed team of oxen and two horses was more speedy than one of oxen only but although this writer possesses an early nineteenth-century engraving showing a horse har-nessed before two oxen in a plough, he is forced to wonder how often, in fact, this procedure was adopted so long ago, or whether the pictures of horse-drawn ploughs of fourteenth-century date show that mixed teams were always unusual.

It is when these works discuss methods of ploughing, ridging up, and so on, that a rumour of classical writing is heard. The directions for preparing the fallow land by three ploughings during the year are much the same as those of the Latin writers, but the source is not mentioned either by Walter, *Seneschaucie*, or *Fleta*, probably because it was, and for ages had been, the established traditional practice of the unlettered peasant. A watchful eye must be kept on the ploughman to see that he did not leave undisturbed land below the furrow, a warning to be found in the Latins. The first ploughing, usually in April, must be done with care, so that the infertile subsoil was not brought to the surface above the good soil. The second ploughing, about midsummer day, ought to be just deep enough to cut off the weeds, and to avoid the land becoming sodden if rainy weather supervened. Little narrow ridges must be ploughed at the third turn so that the seed, broadcast manually, would fall evenly into the furrows, and be covered evenly by the harrow and so cause the crop to come up in a regular manner. It was classical wisdom, too, to sow the seed early enough in the autumn so that it was well forward before the cold weather came. Similarly on clay or strong soil the spring crop must be sown in February to avoid the danger of a hot, dry March, when the soil might become too hard, or too friable or dusty. The seed for winter corn should be got from a distance, and preferably from a different soil, but this was not so important for the spring crops. Walter suggested an experimental proof of this to be made by sowing two selions, one with native and the other with imported seed. He did not normally expect more than six bushels of wheat to be harvested for two bushels sown. This exiguous yield implies a number of problems. Maitland called it the expected product of the high farming of the day, and calculated that if it was obtained at least two acres of land would have to be cultivated to provide the seed for the following year, and a quarter of grain for each adult man. The implication is that the yield may have been more than this, always with great fluctuations, according to the season, but this is not proven. Besides this there was barley for beer, and oats for man and beast. Some wheat, too, may have been brewed. He thinks twenty bushels of barley was brewed per man. The grain made into bread is wisely calculated: whether that made into beer is also I do not know.[71]

Again in the preparation and use of such fertiliser material as could be gathered together there are many resemblances to the advice given in the classical sources though no direct reference is made to them. All the animal waste ought to be collected and kept in a heap. Soil should be added to the heap to increase its volume. Waste fodder, probably straw or stubble, should be put in the cattle yard, or laid in wet, muddy places in the ways. The sheep cote or fold floor ought to be covered with marl, ditch scourings or soil, and straw added. All this material must be gathered together before the dry weather expected in March. It was advisable to use mixed manure and soil on sandy land, otherwise it would prove too hot. If spread at fallowing it would be advantageously turned over at the second ploughing, but spread at the second ploughing it was too deep to be thoroughly intermixed with the top soil when the final preparation of the seedbed was made. Marl was highly thought of: but all sorts of manure were very scarce, and supplies must have been insufficient to provide the requisite amount of plant food.[72] It appears that Marc Bloch was mistaken in saying that marling disappeared after the Roman epoch, and did not reappear until the sixteenth century. Parain calls it the greatest innovation in temperate regions after Roman times.[73] The scanty supply of manure, scanty even when mixed with soil, had to be used carefully, and the contemporary writers were anxious that it should lie close to the seed in the soil, and so be easily assimilated by the plant. Their main tool was the plough supplemented by the harrow, with a few hand tools, but the idea is one that has been experimented with in modern times when more accurate placing of granular fertiliser can be made with machines.

The ploughman was required to help with the harvest, with threshing, digging ditches and cleaning out field drains, in fact he was a general all-round hand. His main duty was to look after his cattle, but the whole is described in some detail, not only in *Fleta* and *Seneschaucie*, but also by Bartholomew Anglicanus.[74] The ploughman's own plaint about his hard lot is familiar in English history books that quote Aelfric's *Colloquies*.[75] The care of the plough team and other farm animals, their feeding and probable yield is dealt with in far greater detail by the thirteenth-century writers than it was by the classical authors, as are the duties of the shepherd, oxherd, etc, but it would be useless repetition to describe all these once more here, because it is ready to hand in many deservedly popular studies.[76]

Fig 5 *Knightly chivalry and the ploughman from 'Moralia' in* Job de Gregoire le Grand *(twelfth century)*

The appearance of these new treatises on farming and estate management is not so surprising as it seems at first glance. For the previous two or three centuries population had been increasing slowly, various economic factors had encouraged the growth of industry and commerce, and consequently more food production was necessary. The aggregation of great estates, ecclesiastical and lay, and the increase of servile status amongst the peasantry, were all elements making for change. All this has been completely described in standard histories, more especially perhaps the expansion of the arable acreage at the expense of forest, moor and marsh.[77] Indeed Henri Pirenne suggested that the expansion that took place until it was halted at the end of the thirteenth century remained at that acreage throughout western Europe until the eighteenth century. It was necessary at the earlier time because farming methods were still extensive.[78]

A great deal of the expansion was the work of the Cistercians and Praemonstratensians whose rules led them into waste places. They reclaimed the wild lands in Yorkshire, England; they helped to drain the Polders; they established themselves in the wilds of Europe. Would it be reckless to suggest that these people were acquainted with the classical methods either directly through the books, or through the traditions of the practising lay brothers and *coloni* who worked for them? Exactly what technical processes these settlers used is difficult to discover. Whether the monks themselves worked manually in the fields with hoe and spade, or only at the harvest is a subject of controversy, but the mere fact that they or the men who worked are said to have carried their tools home from the fields suggests that at first, if not later, the work of reclamation was done in a primitive fashion.[79] Laistner gives a translation of a contemporary Latin verse, 'What great vagary of the muses has led thee astray? Better for thee to hold the plough handle and till the fields, to imitate the husbandman after thy father' example than to sing', which implies that the religious were more inclined to the chantry than the field. It is not, of course, conclusive.

The descent of the literature, scanty as is the evidence, is implicit in all that has so far been said.[80] The books may or may not have exercised much influence in the field, but there had already been technical developments in the practical as opposed to the theoretical world. The latter had begun to look for reasons. The encyclopedists tried to explain plant structure and growth. Walter of Henley had suggested the comparison of yield from two selions planted with local and foreign seed, surely the first suggestion ever made to lay a foundation for practice. Not only was there some basic progress, halting as it may have been, in some twigs—not branches—of the vast body of knowledge we now call agricultural science. There was also the application of the craftsman's ingenuity to harness, weapons, the plough and other tools, to wind and water mills as sources of power.[81]

Thirty years ago Professor Eileen Power was vague on the subject, but believed there was some development of agricultural technique.[82] What it consisted of she did not say! More definition has been given by subsequent writers at the risk of being more precise than the evidence justifies. Some indication of the tendency has already been given in relation to the supposed history of the plough, a subject which will probably never be satisfactorily learned. The literature already devoted to it is vast, and little agreement has been reached among the proponents of the different theories. Whatever else the plough was, it was a roughly made article, mainly wood shaped by the elementary tools of the village carpenter, and with the iron share and coulter beaten out on the primitive anvil of the local smith. Its efficacy in ploughing must have been haphazard. Contemporary pictures show this beyond a shadow of doubt. Nevertheless, whether provided with wheels or not, the medieval plough used in northern Europe was a more complex instrument than the early ard, and doubtless penetrated and inverted some rather heavier types of soil, eg, the loess, than what is normally found in the south.[83]

This plough, a very dubious proposition, is supposed to have dictated the strip system of the open fields early, ie, before Christ, or late, possibly from the eighth to the thirteenth or fourteenth centuries. Maybe the difficulty of turning a long team on the headland, coupled with the proximity of a neighbour's strip, may have done so, but ploughs were not only hauled by horses and oxen, but also by mules and donkeys. In Italy the buffalo was used because it could suffer heat with less distress—in the south—but the ox was most assuredly the most usual tractive power despite the pictures of horse-drawn ploughs that are to be found in small numbers.[84] Besides ploughs a harrow was used maybe fitted with iron, more likely ash teeth. A thorn bush with a tree trunk tied on top was sometimes used. If any farmer had a roller it is unlikely to have been more than a rounded tree trunk. Hand tools, the hoe, mattock, mallet or beetle, were used to complete the comminution of the soil and eradicate weeds. The scythe and the sickle were used to reap the crops, and a jointed flail to thresh some of them while others were trodden out. Already wind and water mills were being used in place of hand mills (querns) to grind the corn. Many of these things had been in use in Roman times.

Some play has been made with an estimated increase in the area under leguminous crops, particularly peas and beans, but these were known to the Romans. An increase in the area under them in proportion to the area under cereals could only be accurately determined by a meticulous examination of contemporary records of cropping. Again an increase in the area under fodder cereals, oats in particular, has been suggested. Similar investigations are required in this connection. There must have been an increase in the area under all crops because of land reclamation.

The additional oats are said to have been used to feed horses used for ploughing, an unlikely development as I hope I have shown. Lucern, lupines, cytisus (tree clover?) were fodder crops known to the Mediterranean peoples. The first was certainly cultivated in Moorish Spain as its Arabic name alfalfa indicates. Lupines and the plant named cytisus seem to have fallen into disuse. The so-called artificial grasses did not reach northern Europe until the fifteenth century.

The time was not yet come, and would not for several centuries, when the theories and advice of the classical writers were applied *in toto* to the fields of western Europe north of the Alps. There were, as there always are, some changes, slight as they may have been: but as John Higgs has recently written 'the open field system was as advanced as the knowledge of the day permitted . . . way of life for all and the best that tradition and experience could devise . . .'[85] The shape of the arable fields was different from that formed by the ancient peoples, but this did not materially affect the processes by which the soil was drained and prepared to receive the seed. The kinds of manure used could as yet be no different from those used in the beginning of crop growing: harvesting was done with the same tools. Not absolutely static, the farming of western Europe remained, in its main features, much what it had been for a thousand years or more. And when it came to writing textbooks describing the best methods known to authors what better authority could there be than the classical texts, or as many as could be laid hands on, for the guidance of the clerks who wished to guide the farmers? No doubt there was room for improvement in the tools used, both hand tools and animal-hauled implements, but these were not to come until later. The horse collar was of benefit to transport; the stirrup, spur and nailed horse-shoe helped the mounted warrior, but did not greatly affect farming.

Progress in that arduous and complex business could only be made when the new and productive fodder crops could be grown to feed a greater number of animals to higher condition, which would make more manure for the cereals, while the leguminous and root crops could be grown on the fallow; and then the requirements of these crops led to the design of new machines to cultivate them.

CHAPTER THREE

The Fourteenth and Fifteenth Centuries

No vernacular textbook on farming was written in western Europe during these two centuries. Anyone who wanted to study the best, or rather the methods recommended by written authority, was forced back on the *Scriptores rei rusticae* if they were to hand, or to the treatises written by Crescentius and the English writers of the thirteenth century. Only one translation of a Roman writer into a vernacular language appears to have been made during this period, that of Palladius into fourteenth-century English verse made for Duke Humphrey of Gloucester.

It is true that the art of printing was discovered in the fifteenth century, and was brought to England by Caxton, but there was no writer on agriculture who took advantage of this new process to inscribe an original farming treatise and get it into circulation. What did happen was that the Roman writers were printed in the language in which they were written. But before discussing the printed editions of these books it will be well to consider how widely they were dispersed in MSS, and who owned the copies. A complete list is now probably impossible to compile, but it is known where some of the copies were. Whether they were consulted for practical purposes remains open to question.

Palladius *de re rustica* seems to have been the most widely distributed of the Latin texts in manuscript form: though any kind of asservation about this must be accepted with the greatest reserve. J. C. Schmitt, who edited a Teubner print in 1898,[1] believed the work was frequently transcribed and read, and the number of manuscripts in being was almost infinite. This is undoubtedly an exaggeration. Although he supplied a list of a large number of manuscripts recorded, it could not by any stretch of imagination be called infinite. Only two years before this edition was published Mark Liddell edited a version of *The Middle English translation* which was printed in Berlin in 1896, but there was no intention here of an examination of the practical value of Palladius. Liddell's edition was produced for the purpose of linguistic study, and he cast some scorn on the earlier edition, oddly written in verse, made for Duke Humphrey. A copy of the manuscript of this translation into Middle English was in Colchester Castle Museum in the 1870s, and a version was printed by the Early English Text Society in 1873.

Duke Humphrey was one of the growing number of princely literati of the later fifteenth century, a patron of Caxton, and a great book collector though somewhat ruthless in his methods. He seized the library at the Louvre to form the foundation of his collection, which he later presented to the University of Oxford. Was this the library, or part of it, accumulated by Charles V of France? He kept some of his books in the tower at the Louvre; others were stored at Melun, Vincennes, St Germain en Laye and Beauté sur Marne. Amongst them was more than one copy of Vegetius in translation, but the catalogues of his library do not mention the Latin text of Crescentius, nor a translation of his book. Le duc de Berry however procured a copy of the original text and a translation. There seems to be no trace of any of the Roman writers in this royal library. Duke Humphrey sent lists of desiderata to agents in Italy which included Pliny the Elder and Varro, so it seems possible that he may have ordered the translation of Palladius for use as a textbook for the improvement of the farming of his estates.[2]

But this was after printing had been invented, though of course people continued to collect manuscripts of classical writings as they have done ever since. Indeed the arrival of new works in Venice, for example, was an event for rejoicing among the *virtuosi* of the early Renaissance, and had been for some long time.

The monastic libraries have been credited with the preservation of the works of the classical agricultural writers. Usually this statement is made in a broad, general, way, and appears to be little more than an inspired guess—if it is inspired! How frequently the monks consulted them if they had them is an open question. Some have been credited with following the textbook precepts on their estates, which are also credited with being generally better cultivated than those of lay lords or peasants. If that is so, these improved systems were not widely, nor even narrowly, imitated. The continual copying of Virgil's *Georgics*, which really contain a scrappy rather than a systematic account of farming, continued as in previous ages, and many of these copies were beautifully illustrated, possibly more as literature than as a didactic work. The treatise of Jean de Brie, *Le bon berger*, was in the same category as Duke Humphrey's translation of Palladius. It was a practical textbook ordered by Charles V of France in 1379, and passed through many editions after it was printed in 1540.[3]

Some doubt had been cast upon the monastic interest over a century ago. It is recognised that many ancient manuscripts were defaced in order to provide parchment on which to copy 'the psalms of a breviary, or the prayers of a missal', and that to some minds of the time reading the classics was an idle occupation.[4] But there was a general elevation of the intellect, not only in the monastic but in the growing university world, and this must have led to the collection of, and commentaries on, texts quite outside the realm of theology. It is perhaps significant that the Carmelites at Florence had none of the rustic authors in their library at the end of the fourteenth century. They did have an Isidore, and other later almost contemporary works. Isidore indeed figured in a good many medieval libraries, but was of trifling aid to anyone wishing to study farming technique.[5]

Other ecclesiastical organisations and eminent churchmen had one or other of the agrarian works. The archbishop of Riga (1304–41) owned a Cato. There was a variety of classical manuscripts at Pisa about 1355, amongst which was one of Palladius. At St Croce, Florence, there was Servius on Virgil, Vegetius and

Lucretius. A Servius was left by Boccaccio to St Spiritus, Florence, amongst other manuscripts, but this is very scrappy. Humphreys asserts that no college (library) had the range of the friars and neither had many agricultural books.[6]

In England during the fourteenth and fifteenth centuries the monasteries were becoming more and more lax and indolent. The cathedral schools with their libraries were more active as were the Franciscans and Dominicans. This is not to say that the books disappeared, but they were no longer assiduously copied, or perhaps looked after. Yet Whethamstede of St Albans had Cato copied. Durham had a Palladius and a Virgil in the late fourteenth or early fifteenth century. St Paul's possessed the *Bucolics* in 1458, and an Isidore. Peterhouse owned a Virgil in 1399. There had been a Palladius at Byland Cistercian Abbey, Yorkshire, in the twelfth century, and another at the Benedictine Abbey of St Augustine, Canterbury. Other copies were at Waltham Augustinian Abbey and Worcester Benedictine Cathedral Priory in the thirteenth century. Moreover, there were three copies of Walter of Henley's *Hosebondrie* in Canterbury Cathedral Library in the thirteenth and fourteenth centuries, and rather surprisingly (to me at least) copies of Crescentius in Hereford Cathedral, York Cathedral Church of St Peter, and also at New College, Oxford in the fifteenth century. Copies of

Fig 6   *Work with a will—when supervised. A simplified reproduction of a fourteenth-century illumination in a psalter*

Virgil, whether of the *Aeneid* or the *Georgics* or *Bucolics* is not disclosed, were scattered about the country.[7]

The devastations of the Hundred Years War did not spare the libraries of France. Jean de Montreuil, chancellor to Charles VI, brought home a Varro *de re rustica* when he returned from his travels in Italy as well as other Latin works till then unknown in France. The monks in the great monasteries of St Gall, Corbie and Fulda seem at this time to have been disastrously ignorant.[8]

Sandys remarks that our version of Varro *de re rustica*, like that of Cato, ultimately depends upon a long lost manuscript formerly in the library of San Marco, Florence, and that Pliny is named nine times in medieval libraries in France and Germany, but only twice in those of Italy and England.[9]

Scholars and collectors were busy in the fourteenth century, especially in Italy. Petrarch owned a Pliny, Varro, Palladius and Virgil's *Bucolics*. The question whether Boccaccio was the first man to possess Cato has been asked, Coluccio Salutato disputing precedence. Columella was known to Petrarch, but the first complete Varro, Martial, and other works were restored to unity by the energy of Boccaccio, or so it is said. Bracciolini (1380–1459) is credited with having discovered some twenty manuscripts of Columella, which he took to Italy. Petrarch tried to get a Hesiod from Constantinople. One Guiglielmo de Pastrengo (? of Verona, fourteenth century) knew Vegetius, Palladius, Pliny and Varro. There were other copies of Columella, Pliny, Varro, Cato, in the private library of the Medicis. Sabbadini's index lists six copies of Cato, seven of Columella, eight of Palladius, fourteen or so of Pliny, eight of Varro, and a good many of Virgil.[10]

All this seems to show that there were only a few copies of the *Scriptores rei rusticae* scattered over western Europe in the fourteenth and fifteenth centuries, a conclusion confirmed by the exhaustive researches of Manitius. In Spain there was none except a couple of printed copies of Virgil's *Bucolics* and one of the *Georgics* in the last years of the fifteenth century.[11] Of manuscripts there seems to be no list.

Though the Greek work compiled by Cassius Bassos for Constantine Porphyrogenitus was translated into Arabic either direct from Greek, or through Syriac or Pehlevi, I have found no mention of any manuscript in western Europe in the catalogues to hand. Kraus said that amongst the Jabirean treatises is one *kitab al filaha* (book of agriculture) in which the majority of agricultural methods mentioned in *kitab al hawass* are repeated. It was composed of almost literal versions of passages from ancient sources. The *Geoponika* was only translated into French in the sixteenth century, and not printed in England in Greek and Latin until the early eighteenth century. To this it may be added that Bolgar asserts that Greek texts were available, presumably manuscripts, in Italy before 1450, including Xenophon *Oeconomica*, six copies of Theophrastus, and seven of Hesiod, but that the evidence for this is of unequal value.[12]

The duplication of books was simplified and made more rapid after the invention of printing. The second half of the fifteenth century saw the production of many classical works, and amongst them the *Scriptores rei rusticae*. Indeed, Augé-Laribé believes that books about farming were eagerly awaited,[13] which is an opinion of substantial probability with no equally substantial basis of proof. Be that as it may, a collection, *Scriptorum rei rusticae veterum Latinorum*, was printed at Venice in 1470, and five other editions before 1500. Two editions, 1494 and 1496, entitled *Opera Agricolationem*, which included Columella, Cato, Varro and Palladius, were issued from Bonn. An Aldine edition of Theophrastus *Enquiry into plants* was printed from an imperfect manuscript at Venice, 1495–8.[14]

How can the effect of the study of these works, if they were studied, upon the actual practice of farming be measured? Were they, perhaps, only read as literature, the wisdom of the ancients, who were reputedly so much wiser than contemporary man? Was practical farming only the result of traditional wisdom handed down verbally from father to son? Since the only medieval farming textbooks were those of Crescentius, Walter of Henley, *Fleta*, the anonymous *Seneschaucie*, and the work of Grosseteste, and these were already a century old, little or no comparison of theory is valid for the fourteenth and fifteenth centuries.

Until the last years of the thirteenth century the western world had been making advances. Population was increasing, large areas of previously uncultivated land were being brought under the plough, and in Italy and the Low Countries industry and trade was expanding. A series of disasters fell upon suffering humanity from the early years of the fourteenth century. Part were natural and no man could control them. Bad weather was one of these. Another was the plague that followed closely on the heels of famine. Social oppression was a human failing, and aroused resentment and revolt amongst the peasantry. Dynastic wars brought terror and destruction in their wake. The so-called Hundred Years War devastated large parts of France and Flanders.

There was a great famine from 1315 to 1317, coupled with dire disease. The Black Death of 1349–51 is said to have reduced the population of Europe by one-third or even a half. There were more outbreaks, not so severe, but bad enough, during the last half of the century, and on into the fifteenth.

In such conditions most men must have been more concerned to preserve their lives than to study ways and means of making improvements in farming. A succession of catastrophes like these caused social changes, some much the same both north and south of the Alps, but not all. Italy was to some degree cut off from northern Europe by difficulties of travel, and its physical structure with a long sea coast led to the development of sea-borne trade as well as trade between the predominantly agricultural and growing industrial areas.

Great estates (*latifundia*) belonging to the nobles were a feature of central Italy where the nobility lived in their castles, and the cities were surrounded by vast empty spaces covered with woods and forests stretching to the mountains, and frequented by bands of robbers. These areas must have had a marked resemblance to Roman times, but in the area of industrial development there was a market for corn. The south supplied the north. Sicily, Sardinia, Apulia sold to Tuscan and Lombard towns.[15]

Even in the north there was some decline in production after 1350, but this was not so marked in Lombardy as in coastal Tuscany, the Roman Campagna, southern Italy and the islands where devastation was widespread and prolonged. Land went back to wild grass and swamp, and settlements vanished. There was soil erosion in the hills, and sedimentation in the lower lands, conditions that prevailed in other Mediterranean countries.

Farm practice was much the same as it had been in Roman times. It has even been said that it was conducted on the lines proposed in the *Scriptores* after an interregnum during which there had been a relapse to more primitive systems. By the fourteenth century, waste had practically disappeared in the most progressive areas—mainly in the north—and changes in technique 'which in places surpassed the Roman practice had been introduced. New crops were being grown, citrus fruits, rice, mulberries. Other innovations included improvements in animal traction', though what these were is not stated, but presumably the horse collar is meant, 'in agricultural implements, particularly the plough' (again what?) 'and the technique of milling corn and oil'.

Some intensification of production especially near the towns is suggested, achieved by a return to the Roman two-course system of dry farming. To meet the rising demand some arable was cropped for two years in succession, which was destructive of fertility where the land was not irrigated, especially in the central mountains and the dry south. In upper Italy the fallow was sown with a catch crop, something the Romans had known all about and recommended in the *Scriptores*. Was this the revival of an ancient practice, or had it in fact been transmitted by tradition, and continued to be done without comment from observers?

Another practice that derived from the ancients was the system of growing crops together on the same piece of land in *campi arborati* or *piantati*. Vines were trained to grow on trees or posts, and, of course, spaced. On the land between and below them grain was cultivated, and throve because it was protected from too much dry heat. The process continued for a very long time. In the south, outside Campania, the crops were normally grown separately, vines, olives and fruit, and large acreages of corn, while there were extensive pastures bare of trees.[16]

The systems of farming, except livestock, practised in Italy, have been described at such length because it is obviously likely that here the methods of the old Romans would still be pursued if not modified by the effluxion of time and the inattention of disheartened peasants. It was in this country, too, where the taste for collecting the ancients first flowered into enthusiasm greater than ever before, where Dante 'incessantly' studied Virgil, Petrarch added fuel to the passion for collecting and Boccaccio's writings and other activities helped the cause along. Great nobles, the Visconti and the Medici, became assiduous in adding priceless manuscripts to their libraries.[17]

The purpose for which these works were studied was not precisely practical. Linguistics played a large part. The study of Greek, never wholly neglected, fascinated the men of the fourteenth and fifteenth centuries. From an extinct world they hoped to derive unprecedented wisdom in morals and philosophy rather than guidance in the management of production, trade and mundane affairs 'to make life here below worthy of a creative God'.[18]

This is not for one moment to suggest that Duke Humphrey, when commissioning a translation of Palladius into the vernacular, did not have a practical purpose in view, but that was far away on the outskirts of civilisation and late in the period. Virgil was generally more or less worshipped, and his works were read and admired all over the West, but Virgil is not a real guide to the practice of farming.[19]

Tangible evidence of the use of the few manuscripts of the classical textbooks in fourteenth- and fifteenth-century England is tenuous, if not completely non-existent. Notable historians have asserted that they were consulted by great landowners, both ecclesiastical and lay. No doubt there were copies to be consulted, and the duplication of the thirteenth-century books, Crescentius, Walter of Henley, etc, provided the opportunity of reading them.[20] But the changes that took place may, on the whole, be said to be the consequences of the disastrous social conditions rather than of a study or reading the small quantity of literature stored in libraries scattered through the country.

New libraries were being established and older ones expanded in England as learning increased. Examples were St Albans built in 1452–3, the Black Monks at Oxford, where the White Friars also possessed one. The Austin Friars built a library in London before 1364. The Grey Friars' library was founded by Dick Whittington. An inventory of the library of the Austin Friars at York was made in 1372. The early fifteenth century has been called the age of library buildings at monasteries and universities.

Some owned no less than 2,000 or 3,000 volumes, but in his list of classic authors, admittedly selective, Ernest A. Savage does not mention any one of the rustic authors,[21] maybe because he was mainly occupied with the contemporary interest in theology, philosophy, and so on.

Whether the books were studied or not there was little change in the actual processes of farming during this 200 years. The drop in population reduced the possibility and necessity for maintaining the expanded arable acreage now known to have been cultivated in the thirteenth century. Deserted villages became a commonplace of the countryside. Farms, for which no tenant could be found, lay vacant on many manors. There had always been large-scale livestock undertakings in the more isolated hill country, and landlords, desperate for a return on their estates, began first to use vacant land for sheep to increase the remunerative production of wool, next to depopulate and enclose for this purpose land which had formerly been arable. Certain physical advantages followed from this action, though the men who were cast out and deprived of their traditional way of life were reduced to the level of paupers or, at best, wage labourers. Still they were able to offer their strength and skill in a seller's market, and some of them benefited accordingly. It is said, too, that more forage crops were grown, beans, peas, lentils, oats, etc, but little of this owes anything to the classical tradition, certainly not, in my opinion, before the age of print.[22]

Climatic and epidemic disasters were as bad, if not worse, in France and Germany as they were in England. The French countryside, too, was devastated in the long years of the Hundred Years War, intermittent though the campaigns were. 'At the end of the Middle Ages', it has been said, 'France was a heap of ruins'. By 1360 cultivation was almost abandoned, and the land was given over to brambles and rushes. Only an occasional farm or vineyard was to be seen. Not a single castle or fortified town remained in the north-east. The population of the whole country had decreased by 50 per cent. The marshes of Poitou had reverted to their pristine state. The woods spread into the fields. Only at the end of the fifteenth century was it possible to recreate productivity here. One-third of the kingdom is estimated to have been reclaimed once more between 1480 and

1510. The fertile valleys of the Central Massif had long been cultivated, often well above the usual limits, but the limestone tracts were little better than deserts. A large part of the Massif was occupied by great pasture lands on which sheep and cattle grazed. Their milk was made into cheese. There were isolated villages in the chestnut woodlands of the Cevennes.[23]

Germany suffered as much as the rest of western Europe from weather, pestilence and war. Eastward expansion ceased, and there was a shrinkage of the area under cereals. The Black Death was as frightful, and warriors in the fifteenth century destroyed many towns. Of course, the demand for grain decreased in proportion to the falling population. Fields and villages had no occupants, and were abandoned to grazing livestock. This implies a relative expansion of animal breeding, both cattle and sheep, despite much of the abandoned land having fallen down to rough grazing, bush, heather, weeds and grass. Animals were imported from Hungary. In some places wine producing was important. The vineyards along the Moselle and Rhine were flourishing in late Roman times. There were orchards and in some places hops were grown. Dye crops were cultivated. This is excessively synoptic, but none of it indicates any reference to the *Scriptores rei rusticae* for guidance, even supposing the books had been conveniently to hand and that the farmers could read them. Nevertheless, it was in Augsburg that Crescentius was first printed in 1471, of whom Fraas rather condescendingly said that he only repeated the maxims of the Roman writers, a statement which is imprecise, but relatively true. Crescentius also used Arabic sources as well as his own observations of the contemporary scene, one of the reasons why his instructions are not invariably applicable to farming in northern Europe.[24]

The peasantry could not be expected to make any improvement in their methods at a time when their number had fallen so sharply, and did not rise again until the very end of the fifteenth and early sixteenth centuries. A great handicap was the bad seasons, to which was added the continual war and banditry. The peasants could only have learned improved systems from the great landowners who were able to study the available text-books, but these people either let their estates piecemeal, or turned to livestock, the majority of the animals being sheep for wool, milk and meat.

Arable farming all over north-western Europe was subject to a fixed routine, a triennial rotation and compulsory cultivation in Germany, France and England. The only changes were made on land favoured by exceptional circumstances; for example, Lombardy near the rich towns, and parts of Spain developed by the Moors by means of irrigation canals. Here the olive, the vine and the mulberry flourished. In Flanders, too, near the towns and on land reclaimed from the sea, though some of this suffered from disastrous floods, farming was not fettered by regulations. There was a comparatively highly developed agriculture, a transition from grain alone to increased attention to fodder crops. Here it is possible that cattle breeding may have increased at the expense of arable farming.[25]

Towards the end of the fifteenth century, recovery was beginning, and printed versions of the *Scriptores rei rusticae* were circulating. These two events were in parallel, and did not necessarily have any intimate connection. Doubtless the books were examined because the ancient wisdom became more and more highly regarded, and was easier to get at. An estimate of the number of volumes in existence, based on what authority I do not know, is that before 1440 less than 100,000 manuscripts existed while by 1500 there were more than 9 million books in print.[26] These figures must include all sorts of works, and what proportion were agricultural texts it is impossible to guess. It is safe to say that more were circulating than ever before, but beyond that vague assumption it is impossible to go. Whether this flood of literature was effective in promoting the advances that undoubtedly took place in the next century or not, it emerges that before the production of vernacular textbooks began the *Scriptores rei rusticae* could be read by those qualified to do so, and that perhaps in a somewhat tenuous way the classical tradition was respected, and could have been followed. No more can be said.

CHAPTER FOUR

The Sixteenth Century

THE invention of printing in the fifteenth century very quickly put the surviving Latin textbooks on farming more freely at the disposal of anyone who was interested and who was sufficiently literate to read them. Whether these printed editions were read as guides to farming or as representative of the ancient learning is a difficult question to answer, but these sources were certainly freely, and somewhat uncritically, used by the writers of the new Latin and vernacular handbooks that were produced in increasing numbers as the sixteenth century advanced.[1]

The fact that the Roman writers were first printed in the original language was no obstacle to the educated man of the day. Education was largely the process of learning Latin as it had been in effect since the foundation of the Church for whose services and other functions it was essential. Latin too was the general means of communication throughout western Christendom. But by the sixteenth century education had escaped from its ecclesiastical bonds, and a growing body of lay instruction was being given simply because the changing conditions of a new sort of society demanded it. Much of this was legal; some of it was technical especially in the financial area, banking, accounts and so on; some of it was purely craftsmanship although the guilds combined manners and moral guidance with instruction in the technique of the craft.

Some of this change in the approach to learning was the result of the changing outlook. Production for the market rather than for immediate consumption in the household or a narrowly restricted locality such as a manor or a village was becoming more ordinary. So as early as the thirteenth century there had been some development of large-scale production of agricultural goods, both cereals and livestock products, particularly wool. The disasters of the fourteenth century hampered this development, but could not prevent it taking place.

From classical times on, Italy had been unable to supply itself with grain. The growth of the city-states necessarily led to the growth of commerce and to seagoing trade with other Mediterranean as well as with Asiatic countries, by way of long established routes. It was this factor that placed Italy in the forefront of what is called the Renaissance, a century before that new birth came to north-western Europe, including the British Isles. The early commercial development of the Low Countries was more remote and was not so speedily affected, although that area had close commercial relations with Burgundy, Spain and Italy.

Italian trade with Byzantium must always, I think, have included the transfer of MSS from the East to the West, and it is well known that there was some transmission of ancient learning (with Arabic additions and commentaries) from the Arabic world to the countries of western Europe. There never was any real interruption of classical studies in the hundreds of years that passed before the so-called New Learning: but the real change was a change in the aims of the student of ancient texts. Medieval man, with learning very largely confined to ecclesiastical institutions, had his mind directed to considerations of advancement in a future life, rather than to understanding the world in which he lived and died. Naturally he, being in the great majority a farmer, could not avoid some contact with and understanding of nature and its phenomena, because that is what he had to deal with, but much of his interpretation was fantastic and mythical. Now with the new angle of appreciation the classics were studied for their attempts to understand and analyse their natural environment, both inanimate and animate, in what has come to be called scientific investigation. This was not altogether unknown in the Middle Ages as is demonstrated by the work of the encyclopedists, the most eminent of whom was Albertus Magnus. There were a good few others like him whose lives spanned the centuries.

It is possible, though perhaps unlikely, that the tentative interest in the external world exhibited by some of the great men of the Middle Ages would have become more intense without the stimulus of the fourteenth- and fifteenth-century collections of MSS and the duplication of texts by way of the printing press. In fact, it was the stimulus of the study of the so-called scientific works produced during the classical era, accompanied by the growth of lay learning, that brought about the change in mental outlook, a change that involved a dilution of the emotion of reverence, an element of the human spirit. The regard for Holy Writ that had absorbed this element of feeling was not abandoned, but it began to be reduced by the rise of devotion to the classical texts that were so carefully studied, and indeed to some degree imitated. This is an inexhaustible subject, and has occupied innumerable scholars, but perhaps enough has been said to show that, at least in the academic world, the *Scriptores rei rusticae* newly to hand in print, must have been highly respected. Technology was no less important during the Renaissance than the Scriptures, which people like Erasmus, Colet and others, studied so minutely, or the artistic work of Michaelangelo or Benvenuto Cellini, not to speak of Copernicus and Galileo in other fields.[2]

It is possibly to labour the obvious if I repeat that one of the historian's greatest trials is the attempt to cast himself into the mental atmosphere of an earlier age. This part of his task is difficult, impossible *in toto*. Only an approximation can be reached. Even so, conclusions are always doubtful. All must be tentative, relying as they do, almost completely, upon inference. One such inference can be drawn from the work of Joachim Camerarius, printed at Nürnberg in 1577, under the title, *De re rustica, opuscula nonnulla, lectu cum jucunda tum utilia, jam partim premium composita, partim edita*. Such an effort would not have been made unless there was some interest justifying it. Of course it is always possible, though unlikely, that Camerarius compiled this work solely for his own satisfaction, but gratifying as that is, scholars rarely work entirely for themselves. There must have been a possible public for the work. It might be called the first bibliography of agriculture, and ambitiously tried to be all-inclusive. Most countries are covered, ancient and modern, and lost books as well as those current are named. Greek and Latin, Arabic, original and translated into Latin, lost Arabic and Hebrew works, Latin authors recently printed, a list that includes herbals, books on horses and dogs, Isidore and other encyclopedists, and then modern books in English, French, German, Italian and Spanish. This comprehensive work did not come out until the beginning of the last quarter of the sixteenth century when there were already several vernacular textbooks on the market.

It is invidious to attempt to allocate priority to contemporary vernacular farming textbooks, partly because the issue is confused by the appearance of translations of the *Geoponika* and of Crescentius, these works being produced as acceptable textbooks equally with the *Scriptores rei rusticae*. Of these two Crescentius was printed first in Augsburg in 1471. Three other Latin prints appeared during the fifteenth century, and at least two more in the sixteenth. Three Italian versions came out in the fifteenth and nine in the sixteenth century. Several French and one German translation were produced in the sixteenth century, another German in 1602. Latin versions of the *Geoponika* were printed in Italy, Switzerland, and France, and translations into French, German and Italian were also made in the sixteenth century.[3]

All these works were old, some even ancient, but they were the foundations used by the contemporary writers of farming textbooks on which to construct their works, and more, not only the foundations but also the general outlines. And there were two other works, of comparatively recent date, besides those already set out, that were printed for the use of practical men perhaps, theorists and authors certainly. One of these was a *Boke of Husbandry*, said to have been 'amongst Bishop Moore's works, in the public library at Cambridge' in 1908. This was a print, how exact I do not know, of Grosseteste's *Rules*. No date or imprint of this edition is mentioned. The other is James Bellot's *The Booke of Thrift* printed in London in 1589, and is a version of Walter of Henley's *Husbandry*. It had been printed long before by Wynken de Worde about 1510, but I have not seen a copy of this. An anonymous pamphlet, *The crafte of graffynge and plantynge of trees* also preceded Fitzherbert by a few years, but obviously was not a complete textbook on agriculture. Both these books are regarded as derivative, owing a large debt to Palladius.[4]

The claim has been made that the Italians were the first to produce vernacular farming treatises, but, while it is impossible not to acknowledge their priority in printing Latin and Greek texts, it was in English that the first textbook was printed. This was, of course, Fitzherbert's *Book of Husbandry* (1523), but even in England it was, as elsewhere in Europe, in the last quarter of the sixteenth century that the larger number of works on farming was printed.

Oddly enough Fitzherbert does not appeal to the classical writers in detail as so many others did. Naturally he was a Latin scholar because that language was the basic element in all education in his time, and his fluency can readily be established by the homilies with which his work is sprinkled; but these are rules of conduct, not of farming. Fitzherbert has naturally been used by all writers of English agricultural history from Ernle on, but he, like Tusser, gives directions to the farmers of his native land, and does not seem to rely upon the classics, or even the thirteenth century exponents, for his rules. They are sufficiently well known not to need expansion here.[5] Fitzherbert was a grazier owning a fairly large herd of cattle, and was, again like Tusser, a strong advocate of enclosed fields. He provided a good deal of veterinary information.

There were several other English writers of farming textbooks in the sixteenth century, but neither texts in the original language, nor translations of the *Scriptores rei rusticae* were printed in England during that century. The only translation that did come out was Gentian Hervet's version of Xenophon's *Treatise of Householde* first in 1534, again in 1554 and 1573, and in the collection *Certain antient tracts* in 1767. It preceded the Italian version produced by Allesandro Piccolimini in 1543, a somewhat surprising circumstance, because it was, rather naturally, in Italy that both texts in the original languages and translations were first printed.

Some idea of the popularity of these works can be gained from

98

a list which does not pretend to be complete. A collection of the Latin writers, Cato, Columella, Varro and Palladius, was printed at Venice in 1472, at Reggio (1482, 1496, 1499), and at Bologna in 1554. At least nine other editions of this collection were issued during the sixteenth century. Cato and Varro were printed together in Paris (1543) and Lyons (1549). Columella was printed in Latin five times; in Italian twice and in French three times during that century. Palladius was published in Latin and Italian. Pietro Marino made a translation in 1528; Sansovino in 1560–1. Varro was printed in the original language. Both Pliny the Elder and Virgil had been well known in the Middle Ages, but industrious translators devoted themselves to these writers, for example, Antonion Mario Negresoli to Virgil in 1543. Troilo Sabino had produced his *Praelictus in Virgilia Georgica* in 1526. Antonio Brucioli translated Pliny in 1543, and this work came out in later editions, but he had been preceded by Landini in 1476. There were others.

Michael Angelo Biondino produced his *Historia della pianti de Theophrasto Libri III* in 1548; Giulio Cesare Seabigiro translated the book again in 1584 following his *Aristotle de Plantis* in 1566. And the two main works of the intermediate period, that is between the fall of Rome and the so-called Revival of Learning, the *Geoponika* and Crescentius, were not neglected. The great Angel Ambrogini (Politian) lectured on Hesiod's *Works and Days*, Virgil's *Eclogues* and *Georgics* and other bucolic poetry, but these lectures are likely to have been literary, though collected under the title *Sylva*.

Another translation that was made by the industrious Italians was of a modern work. The book had been in circulation for half a century before it was turned into Italian by M. Rosco in 1568 under the title *Agricoltura tratto di diversi antichi e moderna Scrittori di lingua Spagnuola* (Venice 1568). It was reprinted in 1577 and perhaps again in 1590. The original was written by Gabriel Alonso de Herrara, and published in 1539, 1584 and 1605 with the title *Libro de Agricoltura*. It followed the pattern of and was partly derived from Crescentius as was perhaps natural. It owed something also to the *Geoponika* and the classical writers although Herrara did not disclose these debts. If the expanded edition published in four volumes as late as 1818–19 is to be believed, the first edition was published in 1513 thus taking precedence in priority over Fitzherbert. But Herrara's work was of a very different calibre from the English book, which was a fairly brief and concise textbook. Herrara's work, like that of Crescentius, was largely derivative, as the Italian version admits.

Spain at that time ruled a number of Italian States, and is said to have had a stultifying effect upon the development of agriculture in that country. The physical difficulties of the farmers in these two lands were not altogether dissimilar although Spain had the worst of it. The abnormal climate and difficult geographical lay-out of Spain presented an almost insoluble problem. The opposition of the sheep breeders of the Mesta and the arable farmers through whose farms the sheep travelled created a social problem as well, just as transhumance did in parts of Italy, notably the Abruzzi. In Spain there were extensive *depoplados* or deserts of hill, moor and salt marsh where nothing would grow. The situation had been complicated by the Moorish conquest, and the introduction of their system of irrigation by the Arabs, whose African experience made them favour transhumance and nomadic sheep breeding similar to their procedures in the land whence they came. They did, however, introduce some sub-tropical fruits and they were great horse breeders. Donkeys and mules of the finest type were bred. The vine, citrus fruits, cotton and sugar, even maize were produced in special places, but the normal succession of cereal crops, known to the Greeks and Romans, was only cultivated with difficulty, except in favoured places. The set-up was a poor one, and the Spaniards, even the nobility, were more concerned with grabbing the wealth from their overseas Empire than with the development of their own estates—where it was possible to develop them. Herrara's book, derivative as it is, demonstrates clearly that in Spain the only change since classical times had been introduced by the Moors. Though the Spaniards possessed a vast overseas market for foodstuffs in their transatlantic empire, their home producers failed to take advantage of it, and no more Spanish treatises on farming were published during this century.[6]

A very different state of affairs existed in Italy in spite of its somewhat similar problems of climate, elevation and slope. The great literary revival there included the writing of a number of farming textbooks almost all of which owed some debt to classical sources, mainly because conditions had remained very much the same, although the general economic situation of Italy was different.

The best known of the Italian vernacular writers are Agostino Gallo and Camillo Tarello, but Luigi Alamanni preceded them with his long poem 'La coltivazione', first printed at Paris in 1546, where a second edition was issued three years later. It was not till 1590 that it was printed at Florence bound with *L'api* of S. Giovanni Rucellai. It fell between two stools. Excellent as its advice was farmers might have preferred, *pace* Tusser, a prose textbook, and the literati a theme, *pace* Virgil, of a different and what might be thought a more poetic character. Gallo's *La dieci giornata della vera agricoltura* . . . appeared in several versions from 1556, ie, with new titles as the seven, thirteen and twenty days. It continued to be reprinted throughout the seventeenth, and again late in the eighteenth century. He used the *Scriptores* as authorities. Tarello, who is well known because he advised a crop rotation, including grasses and legumes, issued *Ricordo d'agricoltura* in 1567, 1577, 1601 and 1629. It is not impossible that his ideas owed something to the classical practice of sometimes taking a catch crop of lupines or vetches after the early (to more northern peoples) harvest had been gathered in Italy. Maybe, too, some farmers had continued to do this as they had carried on with other ancient methods such as growing corn between rows of fruit and other trees, and of growing elms to support vines.

The catalogue grew longer as the sixteenth century progressed. Books in praise of country life and books of instruction how to farm were written by authors anxious to advise on the pleasures and profit to be derived from rural avocations. It would be tedious to name them all, but perhaps Antonino Venuto may be mentioned, if only because Lastri stated that an edition of his work, *L'agricoltura* . . . *Campi, Prata, Orti, Giardini, Viti, Arbori,* etc, was printed at Naples in 1516. No copy of this can be seen in England, although a book by this writer, *De agricoltura opusculum*, the earliest edition of which is in the British Museum, is dated 1537. It was reprinted in Venice in 1560 and 1561, and is in the Rothamsted Library. Other authors treated of single important crops like the silkworms and mulberry tree, olives and vines, but no one dealt specifically with livestock. There may be examples that I am not aware of.[7]

New crops were introduced from overseas, possibly via Spain. Maize, potatoes and tobacco, and tomatoes later. Rice was developed in Lombardy. So long before as the thirteenth century Crescentius had recommended the revival of the ancient practice of growing lupines as green manure, and the cultivation of rape or cole there. The chestnut was a supplement, if not a substitute for cereals. Flax was grown in larger quantities. Already in the sixteenth century it could be said that the increase of fodder crops allowed more cattle to be kept, reducing the number of sheep and pigs in the north, though vast flocks, improved by the Merino introduced by Alfonso of Aragon, continued to be kept in the old way on the *latifundia* of the more southerly part of the country. Farming, of Pistoias, for example, was divided into three levels of altitude, the mountains, the lower hills and the plain, as it had been by Pliny.[8]

Mechanical skill had been increasing for many centuries, but it was addressed rather to the improvement of appliances used in milling, mining and manufactures than to that of the tools of agriculture. Italy has precedence in the invention of the seed drill. Forti described it as the first agricultural machine that had been invented since the plough. The machine was planned by Tadeo Cavallini, and his was the foundation of the design of all future seed drills.[9]

Fig 7   *Hand broadcasting seed in the fourteenth century after a drawing in the Luttrell Psalter*

There was no such flood of vernacular literature in France as there was in Italy. Jehan de Brie's *Le bon berger*, written for Charles V in 1379, was printed about 1540. Crescentius was translated about 1516, and frequently reprinted: so was the *Geoponika*. Columella took on French dress as *Les douze livres . . . des choses rustiques* first in 1552 and again in 1555 and 1556. Augustin Gallo, the Crescentius of the sixteenth century as he has been called, was printed in translation as *Secrets de la vraye agriculture* in 1571 and 1572. Clearly there must have been a demand for agricultural treatises in France, and equally clearly the classical tradition was in the forefront of translator's and reader's minds though to what measure the advice of these writers was applied in the field it is not possible to say in spite of a spate of modern local studies on the subject.

The most important authors of French textbooks, both designed on the pattern of an encyclopedia of country living, were Charles Estienne, an Italian born in Paris, and his son-in-law, Jean Liebault; Bernard Palissy, and at the very end of the century the most famous of them all, Olivier de Serres, As Augé-Laribé suggested there may have been other popular works that were used to destruction, and this is certainly possible but unproven. Amongst such is one, a copy of which is at Rothamsted, by Pierre Belon, *Les remonstrances sur le défaut du labour et culture des plantes* (Paris 1558). The praise of country life, both in prose and poetry, has never ceased to resound since classical times. Nicholas Rapin and Claude Gauchet (1540–1620 ?) continued the process and Dammertin de Goëlle in *Le plaisir des Champs* sketched a picture of rural life and work, a calendar of sorts.[10] Another work filed at Rothamsted is anonymous. It bears the title *Quatre traitez utiles et delectables de l'agriculture* (Paris 1560).

None of these has received an encomium such as R. E. Prothero (Lord Ernle) has bestowed upon Estienne whose *Praedium rusticum* was, he said, the first methodical work on French agriculture. This work is a collection of tracts, published together, first in 1554. Jean Liebault, son-in-law to Estienne, augmented and published a more homogeneous work, on the pattern of Crescentius and the later *Hausväterlitteratur* of Germany. It was *L'agriculture et maison rustique* (Paris 1567). It was an encyclopedia of rural life from choosing the site of the house to keeping poultry and bees, to which Prothero's praise would have been more appropriate. The book must have been popular. Several editions were published before 1600, and it was translated into English by Richard Surflet in 1600, and Gervase Markham a little later; into German in 1579, and Dutch later in the seventeenth century. A translation into Italian was made under the title *L'agricoltura et casa da villa*, published at Venice in 1581, 1591 and 1606. The instructions given are quite appropriate to that country as to others on the north coast of the Mediterranean, and it is curious that it was not translated into Spanish—so far as I know. The calendar of operations to be carried out on the farm are those suited to southern Europe, where, for example, sheep shearing is done earlier than in the north, and where the harvest comes in rather earlier. There is a good deal about the vine, the olive, citrus fruit, chestnut trees and so on, which demonstrates that the book was written without a full appreciation of the different climate and later seasons of the north. For the preparation of the ploughland the usual three ploughings were advised, and the proposed layout of the fields in rectangular plots would have been impossible for the farmers of the open fields to carry out unless they could have agreed to redistribute and enclose their intermixed strips. Lizerand is rather more definite than I should like to be when he said that the matter is simply taken from the Latin agronomes.[11] The derivative character of a large part of the work is, however, unmistakeable. This criticism, too, applies to the even more famous work of Olivier de Serres.

It does not and could not apply to the ideas of Bernard Palissy, because he was practically a self-educated man, and was not familiar with the classics, though the names of some of the more famous authors were, so to speak, household words. In just such a tenuous way he knew Aristotle and Pliny. By the force of his own observation and possibly to some measure of experiment he reached his conclusions, and for agriculture those conclusions were far in advance of contemporary thinking. He became convinced that the food value of manure was the 'salt', part of its composition. This 'salt' was similar to that contained in the plant, originally obtained by the plant from the soil. It was therefore necessary to replace this 'salt', and consequently the manure heap should be made in an enclosed tank provided with a roof to protect it from the rain that would dissolve the salt and wash it away, thus making the residue quite valueless.

From his description of the process the farmers in the Ardennes practised a type of shifting cultivation that was very ancient, even unto prehistoric times. They cut wood in great quantities and piled it on the ground with an air space below it. Turf was cut and put on the top of the wood. The whole was then fired so that the roots of the grasses were consumed to ash. Not exactly paring and burning, but partly a method of clearing ground for cropping. The ash was spread, whereas prehistoric and primitive man only felled the trees, cut the brush and fired it when it was dry enough, sowing their seed in the ash as it lay. The Ardennes people cultivated a piece of land cleared in this way and got fine crops of rye. The area was then left for periods varying between four and sixteen years for the natural vegetation to grow large enough to allow the process to be repeated. Palissy was convinced that it was the salt in the ash that promoted the fine crops of rye. He was also a protagonist of the use of marl as a soil ameliorative, a practice that had been intermittently recommended since the time of Pliny. Palissy was better known in his own day as a manufacturer of fine ceramics than as a geologist or soil scientist, but has been given a great deal of attention in the past 150 years. Sir Hugh Plat recognised the merits of his manurial theories early in the seventeenth century.[12]

Olivier de Serres, Sieur de Pradel, was of a very different stamp. His book, another encyclopedic work, following the pattern set by Crescentius and so often repeated, *Le théâtre d'agriculture et mesnage des champs* was printed at Paris, first in 1600, so that he really belongs to the seventeenth century, especially as the book was reprinted nineteen times before 1675. Demolon stated that it was first printed in 1554, but this I have been unable to confirm. It was evidently popular, partly because its author was one of the nobility, partly because it was welcomed by Henry IV and Sully, both of whom were intensely interested in the development of the natural resources of the country amongst which agriculture ranked high. Henry IV wanted the nobles who had been ruined by the religious wars, and by the inflation, to return to their estates, and improve them by setting an example of good methods to their peasants, and thus maintain themselves in comfort without the support of pensions and places.

Besides producing a work of literary merit still a pleasure to read, the Sieur de Pradel gave an example to other landed gentry of one of themselves who was anxious and willing to improve his estate by the introduction of what may be thought of as novel methods, trusting that their native intelligence would induce them to follow his lead—a rather derisory hope at that time. Improving landlords were more successful in eigteenth-century England.

Olivier de Serres has been called the father of French farming, but it was not his intention to influence the illiterate peasants. His book was designed for the lower ranks of the nobility, who should be, Olivier thought, content to live upon the produce of their smallish estates in a state of 'natural economy'. It was written in the south, and its precepts are advice to farmers in that part of France. Pradel was naturally fertile, and is said to have been cultivated in Roman times. All the more reason for the reliance Olivier placed upon the Roman writers. He had carefully examined and assimilated the advice of Cato, Columella, Varro, Virgil and Pliny. He was also acquainted with Hesiod. In his travels he had noticed Swiss farming, and that which lay on his road to Paris.

Olivier has been lauded rather beyond his considerable merits. He was not a revolutionary, but more in the character of a revivalist. The Roman systems of farming survived the troubles and devastations of war, epidemic and other tribulations of the Dark Ages. It is not too much to presume that in its essentials, but with some lapses, it continued in the Bas Vivarais where Olivier lived and where the predominant system was not very different from that of Italy—naturally enough for reasons of climate, elevation and slope of the land. Cereals were grown on small fields, roughly square or rectangular in shape, where the terrain permitted. The small oxen hauled the *araire*, the light wooden swing plough, as they did until quite recently. Perhaps they still do. De Serres advised fields 200ft long to save

CHI BEN COMINCIA, HÀ LA META DELL' OPRA,
NE SI COMINCIA BEN SE NON DAL CIELO.

Ecco prostrato al Ciel le preci inuio,
Poiche da cominciar l'Opra, e l'Disegno,
Come linea da punto, ha l'huom da Dio.

Fig 8 *A poor peasant praying for a good crop before he begins ploughing. Note the left-side mouldboard made mainly of wood which would turn a left-hand furrow*

the beasts too much fatigue before a rest. Much of the farming was limited to a few years cropping followed by several years of natural grass; nevertheless, Olivier knew of and cultivated the 'artificial grasses', lucern and sainfoin, in rotation with grain crops. Lucern, which was known to the Romans, survived only in Spain by reason of the careful irrigation tillage of the Moors, who used it as horse fodder. It was taken from Spain to France, but it is unlikely that de Serres was either the first or the only man to recognise its value. The crop is also said to have reached Provence from Italy. Either clover, sainfoin or lucern was also known as 'Burgundy grass' before de Serres's book was published, and was recommended by Heresbach and Googe, his translator (to whom I shall return). Augé-Laribé praised de Serres because of his use and recommendation of the ley in the arable course, because he knew of the necessity for an occasional deep ploughing to maintain fertility, and because he made a distinction between exhausting crops and those which added to the productivity of the soil. More, de Serres understood the management of manure and its use. Viticulture, arboriculture, the minor business of keeping poultry and bees, as well as the major business of looking after livestock, are all included as well as butter and cheese making and so on. One admirer has, however, gone a little far in saying that his advice was good three centuries ago, still is (1886), and always will be. New plants from overseas—maize, tobacco, potatoes, etc—were slowly being made the fashion.[13]

Provence suffered from floods in the sixteenth century: so did the Basses Alpes and the Rhône estuary. The soil, too, is said to have been exhausted by *ecobuage*, a kind of paring and burning. The farming in the north-east and Normandy followed the usual pattern of open-field systems. Brittany was a wooded area with small arable fields and grazing in the waste: but this is not the place for a careful study of the practices of widely differing regions. It will be found in Roger Grand and in local studies.[14]

The only other sixteenth-century French work is a book on poultry by Prudent Choyselat, the *Discours Oeconomique*, probably in 1569. It optimistically suggested that it was possible to recruit a fortune lost in the wars by a small investment in poultry, an unlikely prospect, as some optimists found after World War I.

In the sixteenth century only four native German writers dealt with farming and gardening. The earliest of these was the unidentified author of a calendar of garden and field work that appeared about 1530. There was a translation of Crescentius in 1531. The next and perhaps most important writer was Conrad Heresbach, whose *Rei rusticae libri quattuor, universam rusticam . . .* was issued from Cologne in 1570. Martin Grosser produced his *Kurze und gar einfeltige Anleitung zu der Landwirtschaft . . .* in 1590, and another parson, Johann Coler, his *Kalendar* in 1591, his *Oeconomia ruralis et domestica* a trifle later.

The last three of these writers were conversant with and used the classical writers who were by then to be had in printed versions, both in their original languages and in translation. The early *Kalendar* I have not seen, but probably the same thing applies. Translations of some of the French and Italian books were also to hand. Max Guntz included two books by Tobias Moller, *Sommer Feldbau* (1583) and *Winter Feldbau* (1584) which, he said, dealt with arable and vines, the cereals and gardening, but this was largely from an astrological standpoint. He added an anonymous *Pflanzenbuchlein*, which contained a calendar of seasonal work taken from Theophrastus, Pliny and Varro, and mentioned a book on the weather. The important thing, it has been said, is that the Humanists took from the ancients the system of cultivation, the foundation of agriculture, forestry and forest science.

Heresbach was indeed in the main line of Humanists, being a friend of Erasmus, Melancthon and Sturm. He was not himself a farmer, but he knew, because he lived there, the system followed in Nieder Rhin, which was at that time probably the best in Germany, a very likely conclusion since the region was so near the improved farming of the Netherlands. Heresbach was a widely read man as his acquaintance with the leading scholars of the day argues as does the vast list of authorities he supplied as a justification for what he had to say. Fraas underlined the breadth of Heresbach's reading as heavily as his own bibliography. He thought, reasonably enough, that Heresbach may have been influenced by the French and Italians, and no less by the *Geoponika*, and finally by the Arab culture, which had encouraged gardening and horse breeding in Spain. Fraas, too, referred to Ben Achmed of Seville, who wrote on chemistry, natural history and farming, but this writer is unknown to me. Another source was, according to Fraas, ibn al Awam's *Kitab al Felahah*, which he supposed to describe the roots of agriculture amongst a prehistoric people, the Nabatheans, a conclusion which Renan has shown to be untenable, but that is rather beside the point here. Heresbach, so far from denying his debt to the ancients, proudly proclaims it in his remarkable list of sources.

Unfortunately this led him into some bits of foolishness. Few farmers living in his neighbourhood could have been enthusiastic about camel breeding, nor would they have been able to cultivate figs and almonds, perhaps no vine. Other than these aberrations the book is strictly practical. Dr Schröder-Lembke thinks that he was led to write his book by reason of his classical studies, and that he followed classical example and precept. In spite of these somewhat adverse remarks the book was impressive enough to encourage two English translations.[15]

The general advice given by Heresbach is consequently largely that of the classical writers. Grosser recognised that the necessities were different in Germany from those in Italy, and therefore described current practice as he had seen it. Many traditional methods were still in use. The system of preparing the land for arable crops, the cereals in particular, was to give three or preferably four ploughings, eg, standard practice for 1,500 years. Manure must be carefully conserved, spread and ploughed in before the seed was sown, and organic manure should, if possible, be supplemented by marl and other processes of soil mixing. The best manure was pigeon's dung, but Heresbach's classification was first birds', second man's, third beasts'. Aquatic bird droppings must not be used. All this is pure classicism as is the list of equipment required, and the list of crops, though the remark is made that oats were rare in Italy, but were grown in the Apennines and in the Kingdom of Naples. The legumes, beans, lentils, chick peas, vetches, fenugreek are discussed. Medica, called trefoil in France, *grand trefle* or Burgundy grass elsewhere, was recommended. Cytisus, praised by Pliny, was probably a kind of shrub clover (*Trifolium majus*). Heresbach also deals with meadows and pastures, gardens and livestock breeding. A notable point is that the rotations he recommended included rape and roots. On very good soil the fallow should be manured, winter rape sown, followed by wheat, rye with stubble roots, then spring barley. An alternative was to sow codware on the dunged fallow, and proceed as before. Medium soil was fallowed and manured, then wheat or winter barley sown, followed by millet or fennel (*Setaria Italica*), then roots. The three-course rotation included stubble roots, a sign of progressive farming at that time.[16] Unfortunately the progress of German farming was halted by the Peasants' Wars in the sixteenth as it was again by the Thirty Years War in the seventeenth century.

Switzerland had made no such progress. The Swiss continued to follow the methods of cultivation and use the tools which were the result of a fusion of the Gallo-Roman culture and that of the invading Germans. There were the two systems that may be called normal, ie, the two- or three-field system with grazing on the fallow, and a kind of convertible husbandry with a long fallow, the plough only being put into the ground once in two years. Livestock were few in number, and only a few plants were cultivated, the different varieties of cereals, legumes, vines; hemp and flax in some fields. The farm tools were rudimentary, but both the *araire* and the *charrue* were used, although the second required a strong team of oxen. The fields were generally rectangular although long strips developed on alluvial land or as a result of *parzellisation*. Cattle were raised on the best land, especially on the demesne. Donkeys and mules were much more plentiful than horses. Numerous pigs fed on the mast in the forests. Sheep were kept for the usual purposes of supplying wool, meat, milk and cheese.[17] Evidently the new textbooks of Italy, France and Germany, whether printed versions of the classical writers in their original language or translation or vernacular writings, made little impact here.

But this was not the same in England where Heresbach was translated by Barnaby Googe in 1577, Estienne and Liebault by Richard Surflet in 1600, Heresbach again by Gervase Markham in 1631, but this like a still later translation belongs to the seventeenth century. It was these translations that brought the classical tradition to the English literature of farming. There was, unless I am mistaken, no copy of the *Scriptores rei rusticae*, either grouped together or separately, printed in England either in the original language or in translation, during the sixteenth century. Their precepts were, however, communicated by Googe's and Surflet's translations of the German and French books in an indirect way. The exhortations of Walter of Henley, too, were repeated in a print of his work disguised under the title of *The Booke of Thrift, conteyning a perfite order and right maner to profite lands* by James Bellot, in 1589.

Googe added some English names to the list of authorities found in Heresbach. These were Sir Nicholas Malbee, Capt Bingham, John Somer, Nicholas Yetzwert, Mr Fitzherbert, William Lambert, etc, but none of these except, of course, Fitzherbert and Tusser were, so far as I could discover, actually writers.[18] The translation naturally contained the advice of the original about the root crops, turnips and rape as well as the artificial grasses, but there is no concrete evidence that these crops were immediately adopted by English farmers though in some parts of the country the convertible husbandry was normal: but this question has been fully explored in two recent works as well as in an early essay of my own, and does not need expansion here.[19]

The list of sixteenth-century English didactic treatises on branches of farming is completed with the works of Leonard Mascall, who produced three specialised treatises on grafting and planting trees, on poultry and on livestock, and of Reynolds Scot on hops. Prudent Choyselat's book on poultry was translated by 'R.E.', whose name is not known. This last was reprinted by the University of Reading in 1951. Mascall's books are largely derivative, owing much to classical sources, both directly and indirectly, by borrowings from translations or adaptations.

Scot, too, is particular in his references to the classical ideas about soil classification, citing Virgil, Varro, Columella and Pliny, but protesting that he does not rely upon other men's opinions. He could say nothing of Florentinus's proposal to dig a hole and fill it up again to test the fertility of a given soil, a theory that was indeed suggested a good while before Florentinus wrote, he obviously having this from earlier writers. Hops were not, of course, a Roman crop, and Scot relied to a degree on Flemish practice. This was not from written sources because the only book to be had in Holland was a translation of Estienne and Liebault, *De veltbauw ofte landwinninghe*, printed at Amsterdam in 1588.

George Churchey made a translation in 1599 of Janus Dubravius, Bishop of Olmutz's book about making fish ponds, an excellent supplement to the food supply, still appreciated in England though the country no longer subscribed to the Papal Church. This writer referred frequently to classical sources. Sir Hugh Plat wrote suggesting various substitutes for the cereals as bread material in his *Sundrie new and artificial remedies against famine* in 1596, but this is not a farming book and is only included because of its references to the herbals from Dioscorides to the writer's own times. And Thomas Hyll, the writer of the first English gardening book, must not be omitted because he, too, quoted the classics, and used their authority for his advice.

Little more remains to be said, except that it would be pleasant to be able to determine how many people read the books, but this is another more or less unanswerable question. Some indication is supplied by H. S. Bennett, who gives the number printed of various books, all of which came out in comparatively small editions, a few hundred of each. Ordinary editions were, he said, limited to not more than 1,500 copies in order to protect the workmen whose skill must be fully employed.[20] The number of readers was clearly restricted by these trade considerations, and by the cost of the books, although it is obviously possible that each copy may have passed through several hands, or may have been read aloud to an admiring audience.

This pleasant domestic scene is no more than an imaginative suggestion. No definite record remains of a farmer who used either the ancient or contemporary textbooks as his guide. The general lines of sixteenth-century farming all over western Europe were largely traditional, although some innovations were inevitably made. The books, too, were very largely based on ancient and traditional sources rather than on the practice of the field. Two persons are said to have used Virgil, one of them a real man, the other a fictional character. The man was the Elizabethan poet, Samuel Daniel, who became a farmer in his old age. Whether he flourished or not is uncertain 'for though he was well versed in Virgil, his fellow husbandman poet, yet there is more required to make a rich farmer, than only to say his Georgics by heart, and I [Thomas Fuller] question whether his Italian will fit our English husbandry', a caustic but sensible comment. The fictional character was Triptolemus Yelloly, who figures in Sir Walter Scott's novel *The Pirate*.[21]

CHAPTER FIVE

The Seventeenth Century

THE number of books on farming published in almost all the
countries of western Europe rose during the seventeenth century. The exceptions were France and the Netherlands, where
the preponderance of books relating to life in the country was
devoted to gardening and aimed at the nobility and gentry in
the former, and to some extent to the production of vegetables
and herbs in both. This is not to say that no books on flowers,
orchards and gardens were written in Italy, Spain and England.
There were some, but without making a precise count these
were not so numerous as the more agricultural textbooks and
encyclopedias.

Besides the new books those printed in the previous century
most certainly continued to be read, some of them for a very
long time. Amongst them were the works of men who have
remained celebrated to this day, possibly because their books
were either numerous or frequently reprinted, the most likely
factor being the very real utility of the precepts they enunciated.
Another element in the enduring fame of these writers is that of
translation into languages other than that in which they were
originally written, which made them more widely known than
writers whose work was published only in their native tongue.

The authors who achieved the well-merited distinction of
continuing to be read, if reprinting is any guide, were Agostin
Gallo, whose *Le vinti giornate dell' agricoltura e de' piaceri della villa*

(Venice 1573) was printed at least four times in Venice between 1607 and 1622, and had been translated into French as *Secrets de la vraye agriculture* (Paris 1571), another new edition being issued in the following year. Charles Estienne and Jean Liebault's *L'agriculture et maison rustique*, first published in Paris in 1572, and re-issued four times in the sixteenth century, was reprinted seven times during the seventeenth. This book had been translated into English, German, Dutch and Italian during the sixteenth century, and was so popular in Italy that five translations at least were printed in that country before 1677. An equally renowned French writer was Olivier de Serres, Sieur de Pradel (Vivarais) whose book the *Théâtre d'agriculture et mesnage de champs* seems to have been first printed in 1600, although Albert Demlon, *L'évolution scientifique et l'agriculture française* (1946, p 10) states it was first issued in 1554 and had nineteen editions in fifteen years. It was certainly reprinted several times during the seventeenth century to which it would seem more properly to belong.

The Spanish writer, Gabriel Alonso de Herrara, had achieved fame in his own country in the sixteenth century, and in Italy, where translations of his book had appeared in four editions. Both the original work and the Italian version were published again during the seventeenth century, the original twice, the translation once. In Germany, Conrad Heresbach had produced his *Rei rusticae libri quattuor* in 1570, but though it was reprinted in the following year and in 1594, it was not translated into the vernacular, and its use was by so much restricted in its native country. Barnaby Googe translated it into English in 1577, and this translation was reprinted several times before the end of the century, again in 1601 and 1614, and Gervase Markham, that industrious and prolific writer, thought it worth while to produce another English version, with additions, in 1631. Another edition of this book was issued in 1658 with the title *The perfect husbandman*.

Besides the printed books of the previous century and a quarter there were at least two older books that continued in circulation, and there were the *Scriptores rei rusticae* both in their original language and in translation. All of these were doubtless read—for what other purpose are books printed?—if not for their farming axioms, at least because of their traditional authority. Of the immediately medieval books the *Liber ruralium commodorum* of Crescentius was probably the more influential though the *Geoponika* of Cassius Bassos ran it a close second. The latter was printed at least twice in the seventeenth century.[1]

Burckhardt, who died in 1897, has said that 'Rome is everywhere the conscious or tacit premise of our views and thought; if in the essential intellectual points we are now no longer a part of a specific people and country but belong to Western civilisation, this is a consequence of the fact that at one time the world was Roman, universal and that the ancient common culture has passed over into ours'. The fact that the world was once Roman had been even more firmly in the mind's eye of the intellectuals of the seventeenth century than in that of Burckhardt. The belief that the literature of classical times could not be surpassed by the lesser men of that century was firmly held then, a pleasing modesty, lacking in our present day.[2]

The practical value of the Roman textbooks, as well as the widely expressed admiration for country life found in Latin literature, was praised by Abraham Cowley who died in 1667. Despite it all 'in England only the poor take it (farming) up as a profession and can barely live after paying rent'. The landlords were not, at that time, interested in improving their estates, or so he said. And he sought the authority of Columella for his proposal that one college in each university should be devoted to teaching the principles of farming because Columella had complained that everything else was taught in schools. A similar proposal had been made by Samuel Hartlib in 1651.[3] It was natural enough for a poet like Cowley to admire both the classics and country life: he did not attempt to apply his principles himself. Sir Walter Scott's character in *The Pirate*, Triptolemus Yelloly, is a late seventeenth-century farmer in the Shetland Isles, who tries to conduct his farming in accordance with his wide reading in the Latin writers, Columella being his favourite author, although he also indulged in Tusser and other English works.

With the growing interest in science, the knowledge of the world without as well as the world within, the study of the classics had expanded from the confines of learning for its own sake, the expansion of the human spirit by stimulated thought, to the useful precepts that could be gained in order to simplify the 'mechanick' processes on which life depends. Education it seemed to contemporary men should not only be aimed at the production of rounded personality able to cope with the problems of government administration, central and local and other business, but should be extended beyond the confines of the ruling and mercantile classes to those who laboured at the fundamental work of the world producing food for all and the material goods that made life easier and more worth living. The most important productive function which a major part of the population was obliged to perform was food production. Consequently if they were to be educated at all beyond learning to do the actual work in the fields under parental instruction an expansion of formal teaching was required. Both Winstanley and Hartlib thought that the curriculum of the grammar schools was not practical enough for the boys who were trained there. Milton, too, apparently subscribed to this opinion. There was, of course, a very large proportion of the working population who received no literary education at all. None of the protagonists of technical training wished to do away with the study of the classics. They were the fount of all wisdom. Milton, naturally enough, with his consummate learning, in his *Tractate on education* (1642) advocated, and in the personal teaching of his nephews used, the agricultural works of Cato, Varro, Columella, Palladius, and Pliny's *Natural history*. It has, however, been suggested that he was 'trying to fill gaps in his own reading'. This emphasis on the worth of the *Scriptores rei rusticae* and of the Greeks was placed despite the existence of a growing number of textbooks in the vernacular; and that on the need for technical training illustrated the widening split between the humanities and science and technology.[4]

Since the classical authors were so highly recommended as textbooks as well as for mental training, it would be surprising if the writers of vernacular didactic treatises on farming had not drawn upon them to a greater or lesser degree. Preachers and prophets were multiplying in all the countries of western Europe, but not, of course, at a uniform rate everywhere. It would be difficult to draw comparisons because the data, possibly not even all the books, cannot be discovered now. Nevertheless, it may be ventured that the largest number of authors professing to teach farming can be found in Italy and England. There were many reasons for this. Central Europe was more or less devastated by the heroic deeds of the famous warriors who took part in the Thirty Years War from 1618 to 1648. They also drove off and ate the cattle, and when they carried the livestock from one place to another helped to spread contagious diseases. Despite the Civil War, the Interregnum and Restoration, there was no such hiatus in England. In France the efforts of Henry IV and Sully had only a temporary effect—if it could be measured—and the production of farming as opposed to gardening textbooks was negligible. This may have been, and probably was, a sign of stagnation in the processes employed in agriculture in that country, with the possible exception of the north-east, Artois and Picardy, on the borders of the Low Countries.

The descent of the classical precepts should, it may be thought, be more readily traceable in the writings of the Italian scribes and

Fig 9   *Threshing with the flail, from Crescentius,* De omnibus Agriculturae Partibus (*1584*)

preceptors than in those composed in countries more remote from the centre of the Roman state. This is all the more likely because there were traces of the processes employed 3,000 years ago to be found in Italian farming until quite recently: but since the whole of western Europe enjoyed an education that was basically classical these authorities were used everywhere—most frequently in the more numerous publications of England and Italy.

The translations into English from the Latin of Heresbach and the French of Estienne and Liebault, the former both by Googe and Markham as already mentioned, and the latter by Richard Surflet in 1600 and again by Gervase Markham, augmented he claimed, in 1616, were obviously vehicles of the classical tradition. The original works contained many references to ancient wisdom, although some of it was certainly not suitable for application to north European conditions. The same can be said of the other sixteenth-century encyclopedic works, and of the translations of Crescentius into French, German and Italian. It would perhaps be a work of supererogation to make minute comparisons between these texts and the *Scriptores*, desirable as it is to do this between the currently, seventeenth-century, published works in order to give some weight to whatever derivative debts they owe.

Gervase Markham has no peer as a writer of farming text-books, both original and translations, at least in the volume of his output, some of which was unashamedly repetitive, some of it as absurd as parts of the works of other writers, some of it eminently practical. His reputation was breathed upon in his lifetime, but he has recently found two protagonists, and older opinion must therefore be somewhat revised. Indeed, Donald McDonald had a good word to say for him so long ago as 1908, slightly qualified by the old criticism, but his brief biography pointed out that Markham was an excellent classical scholar.[5] His lists of authorities consulted amply confirms this, and his acquaintance with then modern foreign languages also. Some of these authorities are, it must be admitted, difficult to identify. It is not altogether surprising that Markham commenced authorship by writing books about horses, from which he proceeded to farriery as veterinary medicine was then known, but this is a subject that can only be touched upon here. His books on general farming were comprehensive, and, as they continued to be reprinted throughout the seventeenth century, must have been popular and useful.

Two other writers whose books were reprinted during this century were Leonard Mascall and Sir Hugh Plat. Mascall is frankly derivative, but he claimed to have made experiments with orchard trees. The book on planting and grafting 'all sortes of Trees' is acknowledged to be 'By one of the Abbey of St Vincent, France' with some additions from Dutch practice. His *Government of cattle* was repeatedly reprinted, and finally much enlarged by Rich. Ruscam, Gent, in 1680. This book owed a heavy debt to the classical writers. The original preface mentioned Plato, Cicero, Aristotle, Virgil, and throughout the book frequent use is made of Varro, Democritus, Mago, but where he got the last cannot now be known, Pliny, Dioscorides, Celsus, Columella. Not only these authorities indicate the continuing influence of traditional wisdom, but it is underlined by the citation of 'Stephanus', who himself relied upon the *Scriptores* about cattle keeping. Fitzherbert's advice on buying and the Dutch method of housing beasts and feeding boiled turnips cannot be similarly attributed.

Another book that was frequently reprinted after its first publication in 1594 was Sir Hugh Plat's *The jewel house of art and nature*. This owes no debt to the classics, but much to the Frenchman, Bernard Palissy, whose theories about plant nutrition and the use of manures it brought to the notice of the English public, a wasted effort because the theories advanced were ignored for a very long time. It is included here because it is indicative of some sort of breakaway from the classical tradition, and the small beginning of independent and scientific, or pseudo-scientific thinking. Already there was a growing body of opinion that the process of acquiring natural knowledge should be the careful examination of the physical environment, if only as a supplement to the ancients. The great names of the time, Gilbert, Bacon, John Wilkins, Robert Boyle and many others are significant enough without going more deeply into this subject.[6] Both Gabriel Plattes and his patron, Samuel Hartlib, emphasised the point in terms that are not very explicit to a modern reader.

The two most important Englishmen who wrote on farming in the second half of the seventeenth century were Walter Blith and John Worledge. There were other writers, some general, and some who dealt with specific subjects, and some pamphleteers as well as horticultural authors. Blith devoted himself to currently possible improvements such as water meadows, draining fens, enclosure, and so on. He seems to have been a very religious man, and quoted freely from Holy Writ to support his moral maxims, but made no specific reference to the classical writers, unlikely as it is that he was not acquainted with them. Worledge, who invariably founded his advice, or at least supported it, by reference to earlier authorities, made it quite clear that he was widely read in Latin literature, and cited every English writer from Fitzherbert and Tusser to his immediate predecessors.

Although the restored aristocracy, especially the lesser lights, had more interest in making their estates pay, often by farming themselves, there was no great spate of farming textbooks between 1660 and 1700 in England. There were some others besides Blith and Worledge, but they are not of outstanding importance at a time when physical innovations played a greater part in life than writing or reading books about how to make them. The end of the century was distinguished by the issue of a weekly periodical dealing with husbandry and trade under the guidance of John Houghton.

The number of original textbooks on farming and country life published in Italy during the seventeenth century was not perhaps so large as in England, but some ideas of almost startling originality were set out. These books and the translations from other languages cannot have had a very great influence on the systems followed by the peasantry, if the somewhat sweeping assertion made by Bandini can be accepted. Until the eighteenth century, he wrote, Italian farming in all aspects was very little different from what it was in the Roman epoch.[7] This idea receives some support from the observations made by travellers, Bishop Burnet at the end of the seventeenth century, and indeed from others at a much later date. Burnet's general impression was that the whole country was almost dispeopled and very miserable, but there were some redeeming features. Lombardy was laid out in fields of one or two acres enclosed by trees. The soil was most desirable, and the work manually intensive, as intensive indeed as the prevailing poverty. Parmesan cheese was made at Lodi. Other parts were cultivated as intensively, probably mainly by manual processes, although he does not say so, as a garden. This was so from Milan to Brescia, and from Vincenza eighteen miles to Padua. About Bologna the soil was cultivated with all due care. Naples was then the richest part of Italy, producing wine and oil. Apulia was a great corn country. The Grisons and Valteline produced fine meat and poultry from the summer grazing; but in strongest contrast large areas of the country were desolate and uncultivated. There were vast heaths seven or eight miles long before Verona though beginning to be cultivated. The Venetian peasants were taxed to poverty. The Ferrarese was abandoned and uncultivated because of neglect of drainage, which made it so unhealthy as to be uninhabitable. Similarly the Campagna was deserted. In many places in Tuscany the soil was quite neglected for want of hands. From Pont Centino, Monte Fiascone to Viterbo, a vast champion country was almost deserted. From Rome towards Naples, in an area 100 miles long by 12–20 wide, there were very few houses, it being very unhealthy because the drainage was neglected.[8]

This somewhat desolate picture is modified slightly by the observations of English travellers earlier in the century: but not to any great extent. Moryson remarked upon the economical use of land in fertile Lombardy and Venetia where corn was sown between lines of olive and almond trees, and where elms were grown to support vines, a system inherited from the Romans and doubtless practised through the centuries. In the Botanical Garden at Padua, Coryate found a plane tree, a species then unknown in Britain. Sandys noticed the silk culture at Messina. This wretched state of affairs must have stimulated the production of farming literature intended but, owing to their economic situation, unable to stir the sluggish peasants into more formidable activity.

The stagnation of rural life was quite opposite to the intellectual activity of the country—in urban circles. Evelyn praised the Accademia Secretorum Naturae established in Naples in 1560 as the first (earliest) scientific society in Europe. The Accademica del Cimento was founded at Florence about 1655. Similar societies were formed in France at the beginning of the seventeenth century, and there were others in Italy.[9]

Two events of importance happened in Italy at the turn of the sixteenth and seventeenth centuries. One was technical, the other perhaps more in the nature of advice that was not necessarily closely followed. The technical event was the design of a seed drill by Cavallini, a seed drill that embodied all the essential mechanical arrangements found in the modern machine. It did not have much of a material success, and the idea of such a machine was discussed in other countries for two centuries or more before it became an ordinary sight in the fields.[10] The diffusion of a theory was made by the publication of Camillo Tarello's *Ricordo d'agricoltura* first in the late sixteenth century and in 1601 and 1629. The important part of this work is the suggestion that the fallow should be abolished in favour of a sort of four-course rotation to include some of the artificial grasses like lucern, clover and sainfoin, which played so great a part in the development of later farming in northern European countries. Whether this idea was derived from classical sources or not is open to question. Tarello quoted Columella, Pliny, Virgil, the *Geoponika*, and Crescentius in various connections. The crops were certainly known to the Romans. They continued to be grown in Moorish Spain, where lucern may have been re-introduced as alfalfa. It was known in southern France, at least to Olivier de Serres, and is believed to have reached northern Europe via Burgundy or from Italy across the Alps to the Netherlands, whence England learned. The process can be described as a result of the classical tradition either in a practical or a literary way or both. It was a proposal rather scornfully remarked upon by Vincenzo Tanara who said some moderns had other opinions, amongst them Camillo Tarelli di Lunato, and these people wanted fields disposed in four parts to permit a four-course rotation to be followed: but Tanara's book is printed in italic, small and close in a very compressed way, and is almost impossible to read. A third edition of it was published at Bologna in 1651 with the title *L'economica del citadino in villa*. It owes a debt to the classics or was perhaps merely repeating traditional peasant wisdom when making the suggestion that if there was little manure to hand it would be well to grow lupines for ploughing in. His ideas about soils are completely classical. Sandy land is the worst: chalk is not good; the natural ecology is an indication of quality. Oak, for example, denoted good soils; chestnut trees, inferior. This sort of thing is pure classicism.

Similar dependence upon authority is found in P. D. Vitale Magazzini, who produced an agricultural calendar in 1634 with the rather resounding title *Coltivazione Toscana*. Not only is some of the matter derived from Crescentius, but Palladius, Pliny and Columella are referred to. This book like so many others written both by Italians and by persons of other nationalities, placed more emphasis on a sort of garden culture that must have been practised by the ancient Romans, who only farmed two jugera or even later when this size of holding had been somewhat enlarged. Unknown to old Rome, the crops brought to Europe from America are discussed, amongst which were the potato, maize and tobacco, the last of which commanded a treatise all its own in 1669. It was by Benedetto Stella and its title, *Il tobacco . . .* [origin, history, preparation and quality] *in fumo, in polvere, in foglia, in lambitivo et in medicina, etc* (Rome). Less esoteric were the books about special crops, the vine, the olive, the mulberry tree, and the silkworm. On the first two subjects there was little new that could be said. Good wine and oil had been produced by very much the same methods ever since these had been cultivated.

The plan of many of the agricultural books written in all the western countries during the seventeenth century was very much alike. The opening chapter or chapters dealt with the choice of site, some with the layout and construction of the house and buildings, and then proceeded to discuss crops, and stock, housekeeping, herbs for medicinal and culinary purposes, the care of fruit trees, vines, etc. This must already be apparent. Gio Battista Barpo followed the regulation procedure in some degree. His book, *Le delitie & i frutti dell' agricoltura e della villa . . . lib III* (Venice 1634) opens with advice on the choice of site, etc, proceeds to animals, poultry and pigeons, then goes on to arable, placing some emphasis on the need for manure on cold land, and a long fallow for worked-out land. Soil-improving crops were beans, millet, lupines, rape, cabbage, etc, wise advice partly if not wholly derivative. Barpo referred to Clemente Alessandrino as his authority that water which had passed through sewers was good for grassland. There is some difference in name, but this Clemente whose book continued to be reprinted until 1696 was probably the same man as Clemente Africo who published *Dell'agricoltura accomodata all'uso dei nostri tempi . . . lib VI* (Venice 1572) a work that owes a great deal to the *Scriptores* but also sets out the writer's own experience. Book II of Barpo's work discussed gardens and vines, subjects that provided ample opportunity for reference to Horace, Martial and Virgil. Book III dealt with surveying.

Gio Marcia Bonardo repeated much of the advice found in the *Scriptores* in a reprint of his book *Le richezza dell'agricoltura* that came out in 1654 some seventy years and several editions after its first appearance in Venice in 1584. It is another example of the continued use of technical works already nearly a century old, some of them more, in which only a modicum of new suggestions was made because the observed practice of farming which the authors saw around them was little changed from ancient times. One other late seventeenth-century writer is worthy of inclusion because he claimed that his book was animated by reason and experience and was not the result of reading old authors. He was Giuseppe Nenci, and his book was *Reflessi sopra le piu frequenti e necessari operazioni della coltivazioni* (Florence 1691). His advice is not very, if any, different from that of his predecessors. What else could he recommend? A point of interest is perhaps that he said that the spade and the hoe were to be used in making the autumn seedbed, probably on very small farms, the plough being used on larger ones.

No more than in relation to English books does this examination pretend to have discussed the total output of Italian agricultural writing of the seventeenth century. From this somewhat cursory discussion it emerges fairly clearly that the precepts of the *Scriptores rei rusticae* were not only respected if known about, but were also the basis of practice, possibly somewhat deteriorated in parts of the country, and a little improved upon in others, except in favoured localities like Piedmont, Lombardy and Tuscany, and only in some places there. Therefore, both in theory and in practice, Italian farming in the seventeenth century remained much the same as it had always been, the suggestion of Tarello for improved rotations and the mechanical ingenuity of Cavallini notwithstanding, but it is inappropriate to the present purpose to say more here. Many modern historians have been busy about Italian agricultural history, and their writings are very extensive.

The great man of French agriculture in the seventeenth century was Olivier de Serres, just as Herrara was the foremost Spanish agricultural writer in the previous century. In Spain the farming literature that followed Herrara was slight, only a couple of books being printed. One of these was a translation of Aristotle entitled *Historia general de aves y animales*, made by Diego de Funes y Mendoza in 1621, and an anonymous *Agricultura practica* printed at Barcelona in 1626. In France the development of farming textbooks was almost equally negligible, but the popularity of de Serres compensated for the lack of competitors.

On the title page of his adaptation of Surflet's translation of Estienne and Liebault's *Maison rustique*, Gervase Markham declared that he had used several authors to expand and improve his version. The named writers were 'Serres his agriculture, Vinet his Maison champestre [French], Aleyterio in Spanish and Grilli in Italian.' Of these the French writers are easily identified. The first has been mentioned and will be discussed. The second was a joint production of Elie Vinet of Saintoigne and Antoine Mizauld de Molluson entitled *La maison champestre et agriculture*, in five parts, 1602. The title is a trifle misleading. The contents are devoted to surveying and measuring land; laying out gardens; garden culture; herb garden medicinal; vegetables and fruit trees and the vine. No doubt this was of service to Markham when making his translation because it covers some of the same subjects, though it does not deal with crops and livestock. The Spanish and Italian writers cannot be identified—at least by me.

To return to de Serres. His sources are the *Scriptores rei rusticae*, and this is emphasised by the nineteenth- and twentieth-century writers who revived his memory. Indeed the Bibliothèque Nationale celebrated the fourth centenary of his birth by a formidable publication, etc, proclaiming that he did not write for peasants who could not read, but for the seigneurs and large landowners. The *Thétâre*, one writer declared, was ordered by Henry IV as much as the *Georgics* by Maecenas. De Serres's favourite literature, which he continually read, wrote Edmond Pilon, was Hesiod, Cato, Varro, Columella and other ancient authors who wrote about country life. Virgil he read at night, or so Pilon said, but the reason for this somewhat questionable statement does not appear. In 1879 Henri Vaschalde said that Arthur Young, when he visited Pradel ninety years before, looked upon the place and burst into tears, a touching story indeed. This critic made the portentous statement that de Serres's advice was good three centuries before, still is, and always will be. Who would be a prophet? The theories and practices de Serres advanced were far outside the experience of the peasants who only carried on as father advised son with the hoe in his hand. Farming followed tradition, and was not embarrassed by theories, a rather cynical remark by Marc Bloch in the memorial booklet. Charles Parain in the same publication underlined de Serres's extensive erudition, and after a learned peroration on the classics, the *Geoponika* and Crescentius, mentioned the *Maison rustique* and Bernard Palissy, especially on manures.[11]

Fig 10   *Orchard work in the reign of James I, from William Lawson,* A new orchard and garden *(1618)*

De Serres himself is proud of his references to the ancients, a foible that is quite easily forgivable because his own estate is reported to have been fruitful in the time of the Romans, and the general farming of the south of France, like that of Italy, had not been greatly modified since old times. He said that the ancients supplied rules, but that there was no rule that was better than experience. In one respect at least de Serres was careful. Cato had advised farmers not to change the shape of the plough 'soc', and to suspect all novelties, a warning that most farmers, particularly those who were illiterate and could not read, would have found quite unnecessary.

To select a random example amongst many that might be cited, de Serres advised that lupines ploughed in as green manure had been approved by the ancients. In his day this practice still prevailed in modern Tuscany and Piedmont where the plant was sown in June or July and ploughed in before sowing wheat. Since the legumes played so large a part in the development of modern agriculture his approval of them is praiseworthy. De Serres was perhaps no more responsible than say Estienne and Liebault for praising these plants, and he is no more certain than other writers about the botanical differences between them, but his influence must have been of value. Sainfoin was, he wrote, called sainfoin in France, medica in Provence and lucern in Languedoc. The ancients had valued it more than other grasses. It was little known in his day in Italy and Piedmont, but well known in Spain and France, principally in Languedoc, Dauphiné, Provence and thereabouts. It could be cut five or six times a year, occasionally eight or nine, an optimistic estimate that was re-peated by the self-styled cognoscenti in other countries. In many other matters of de Serres's work the same dependence on classical authority is evident, but too many detailed comparisons could easily become tedious.[12]

A book, *La nouvelle agriculture* . . ., translated from the Latin of Pierre de Quiqueron, Bishop of Senes, by François de Claret, was printed at Arles in 1613, and re-issued in other editions. The title is promising but the contents rather disappointing because the promise of a treatise on novel farming systems is not fulfilled. It refers readers to the *Maison rustique* and other dictionaries for details of species of cereals. On other branches it cites Pliny, Columella and Termillius Pollio, and discusses Cicero at some length. Although it mentions most livestock and the semi-tropical fruits, figs, prunes, apricots and rice, saffron, etc, it is not really technical. The praise of Provence is marked, especially the horses bred in that area and in the Camargue. De Musset-Pathay said it was first printed, presumably in Latin, at Paris in 1551, and condemned it as undigested and confused, but useful as a guide to the farming systems of the time, rather high praise that cannot be substantiated. This work has adopted even more from the classical writers than de Serres.[13] Yet another rarely mentioned work was by Loys Guyon, Dolois, Sieur de la Nauche (Limousin) entitled *Les divers leçons* (1604), which the writer admitted was collected from good authors, Greek, Latin, Italian and others. Maize, called Turkish wheat, is dealt with and ways of improving land without manure are detailed though the author laid stress upon the value of animal droppings, and the special value of pigeon dung, evidently on classical authority. He called upon Hesiod for support in recommending ploughing in the haulm of lupines, chick peas, beans, vetches, lentils, and similar plants.

Theodor von der Goltz thought that in seventeenth-century Germany the Roman writers were still considered the best sources of information about farming problems. This emerges pretty clearly from the contemporary vernacular literature on the subject, and there was, by virtue of the printing press, some carry over of books written in the previous century, but the formidable destruction conducted by the heroes of the Thirty Years War caused a break in the output of German farming treatises between 1618 and 1648, and indeed somewhat longer. No more than in any other country were the books of any service to the peasantry, who were not only illiterate, but were, Albion W. Small pronounced fifty years ago, regarded by the wisest men of the time as incapable of successful initiative 'and therefore fit for action only under authority', the ruler and the government being in some ways their guardians. The books were written by educated men, parsons, landowners, and so on.[14]

The best known of the early German agricultural writers, after Heresbach, is Johann Coler, who published a calendar of monthly works in 1591, but whose main work was *Oeconomia ruralis et domesticus*, first printed in 1593. No fewer than fourteen editions of this work were published in less than a century, ie, by 1711. It was another compendious treatise along the lines of Crescentius, Estienne and Liebault, and Olivier de Serres. Not only was it a valuable farming textbook, but it covered the many and varied branches of housekeeping, and was thus perhaps the begetter of many imitators, the *Hausväterlitteratur* of the seventeenth and eighteenth centuries. Being a parson, like his father, Coler was acquainted with the classical writers and quoted them frequently. He even quoted Ausonius. He had read Camerarius, Crescentius, and Heresbach as well. When writing about arable farming he more particularly sought authority for his statements from the classics, his list of whom (p 89) is staggering: Aristotle, Xenophon, Virgil, Varro, Columella, Theophrastus, Hesiod, Palladius, Pliny, the later *Geoponika*, Robertus Britannicus, Albertus Lollius and many others, or so he says. And he had a good deal of experience of local practice as any parson who works in a rural benefice must. His book may therefore be taken as a description of the agricultural and social conditions in Mecklenburg about 1600. Here and there it is marred by a tincture of astrology.

Oddly enough another *Oeconomica* was printed in the same year as Coler's book. This was from a manuscript written by Abraham von Thumbshirn, *Domänen verwalter des Kurfursten August von Sachsen*. The title of the book recently reprinted under the editorship of Dr Gertrud Schröder-Lembke, 1965, is *Oeconomica oder notwendiger Unterricht und Anleitung wie eine ganze Haushaltung . . . kan angestellet* (Leipzig 1616). The author was of a different stripe from Coler and others. It is possible that he was a literate practical man more concerned with the problems of estate management and directing farm operations rather than with the study of the classics as most parsons and noblemen were, a suggestion that may perhaps explain the almost complete lack of reference to classical authority to be found in his work. Fraas compared the book with the *Capitulare de villis* of the ninth century.[15]

An oddity worth mention was Moller Albin's *Der grosse alte Schreitkalendar*, 1605, though it cannot be described as a farming calendar. It discussed the kind of weather that might be expected each month, and supplied a blank page opposite, presumably for notes of the actual weather for comparative purposes—a speculative observation. Most of the matter is astrological. This calendar was evidently published annually for some years.

No new work was printed for many years, but the value of Coler's book was recognised by the printing of new editions during the difficult times of the Thirty Years War. After 1648 frequent examples of *Hausväterlitteratur* appeared and continued to do so well into the eighteenth century. These were encyclopedias of rural living for the use of the upper classes, like the Italian and French works of this kind. Consequently they reflect the conditions in which these classes lived—not of the peasantry. Their value as sources of agricultural history varies widely. The accepted opinion is that the best of them is *Georgica curiosa oder adeliches Landleben* by Wolff Helmhard, Freiherr von Hohberg, published at Nürnberg in two volumes in 1682. In Dr Schröder-Lembke's opinion, von Hohberg, though a learned and widely travelled man, was not personally acquainted with the details of practical farming, and so had to rely on the literature and the experience and observations of some of his contemporaries. This judgement is well grounded because von Hohberg's own list of authorities is almost all-inclusive. Reasonably, Coler appears. Earlier works named are the *Geoponika* of which von Hohberg remarked; Latin, German and other translations; Crescentius, whose book had been translated into German in 1602; most of the Italian writers described above; Heresbach; Estienne's *Praedium rusticum*, and Estienne and Liebault's *Maison rustique*, translated by Sebizio into German; Olivier de Serres, the 1635 edition; Herrara, translated by Mambrino Rosco da Fabriano into German, a book in which von Hohberg found many references to Greek and Roman writers; Agostino Gallo; but to complete his list would be to make an almost complete bibliography of the then extant farming literature, including the classical writers. It even mentions Worledge's *Systema agriculturae* with a list of that book's contents.

Such an array of authorities suggests that von Hohberg's book, a massive two-volume folio, was intended to supply an international conspectus of contemporary agriculture, but this it does not do though its author was well aware of the newly developing forms of intensive farming in western Europe, including parts of western Germany and of Italy.[16]

Several other books fall into the category of *Hausväterlitteratur* but, of those printed in the seventeenth century, Dr Schröder-Lembke mentioned only one, and Professor Abel none, though both record the early eighteenth-century work of Francisci

Page 133 *A ploughing scene from* La vie et des miracles de Notre Dame *de Jean* Niélot (*late fifteenth century*)

Page 134 *The Harvest, from L. Liger (1721)*

Phillippi Florini (Pfalzgraf Franz, Phillip bei Rhein), *Oeconomus prudens et legalis*, which is not precisely an agricultural work though partly based on von Hohberg and other earlier authors, and therefore some classical dependence, if only indirect. Another was Joh Chr Thieme, *Haus-Feld-Arney, Koch, Kunst-und Wunderbuch* (1682). Von Hohberg named others, and Fraas has a longer list, including Boeckler's *Haus und Feldschule* (1666, 1683, 1699), Chr Hermann, *Haushaltungsbuch* (1674, 1677), von Hering, *Oekonomischer Wegweizer* (1680), Fischer, *Haus*, etc (1696), J. Jac Agricola, *Schauplatz* (1676, 1678), and there were other writers in this genre, but none equal to von Hohberg in utility and breadth of learning. Most of the authors were prudent men who avoided novelty, and confined themselves to repeating what had often been said before, sometimes in the exact words of their predecessors, ancient and modern. The classical tradition was thus emphatically carried on, at least in literary agricultural circles, although some writers had already recognised that precepts of value in Mediterranean countries were not exactly suited to the harsher climates, and possibly more varied soils of north-western Europe.[17]

There were other books dealing with horse breeding, veterinary notions (not yet a science) relating to horses in the main, and with gardens for pleasure and profit, but these ought to be dealt with separately by specialists in these particular subjects wherein I am but a novice. The practical world must have gone about its business with no very great interest in all this writing and scholarship. The devastation caused by the wars had to be made good by recolonisation, but this was accompanied by the constant threat of starvation if poor harvests supervened as they often did all over western Europe in the seventeenth century. The main development in the processes of farming was made in the Netherlands, as it had been about the same time in parts of northern Italy and perhaps in the Mediterranean provinces of France. It was from Brabant and Flanders that improved farming was brought to England and later to Germany.[18]

Although the early progress of Netherlands agriculture had so marked an effect on farming in other countries no truly husbandry textbook was written in that part of Europe during the seventeenth century, this defect being partly compensated by Sir Richard Weston's description and the publication of Hartlib's plagiarism entitled 'his Legacie' in England. A few horticultural works were all that was published in the Netherlands language, and these had little or no reference to the classics. The excuse for this paucity of literature at a time when writers in other countries dealt so exhaustively with rural living and farming technique was that the Netherlanders preferred to farm rather than to write about it. Clearly therefore any tradition of the precepts and rules laid down by the classical authors must have been vocal, handed down from father to son in that region, if indeed it ever existed. The rural conditions in the Low Countries were so different from those in early Italy, or elsewhere on the Mediterranean, that the wisdom of the ancients would perhaps have been of little service there, except in so far as it concerned the fodder crops that were cultivated by these people at a very early date.

This indifference to the classics was in some degree paralleled by the impact of what, for lack of a more precise word, may be called the development of science in the seventeenth century. Already in the sixteenth century Bernard Palissy had promulgated some suggestive ideas about soil science and the relation between the composition of the soil and that of the plants it supported. His ideas were, if not absolutely original, at least independent of a classical training. It would be extravagant to suggest that the German, Johann Rudolf Glauber, 1604–88, was a follower of or successor to Palissy, but he also believed in the relation between what he called salts in the soil and the plant as an explanation of the production of vegetation. This is excessively simplified, but indicates a breakaway, a new outlook, not determined by a classical education, although Glauber undoubtedly knew the classics. He could write in Latin like many other educated people of his time.

The strength of this independent thinking should not be exaggerated, at least in relation to farming theories, but it was certainly present as the volume of literature on the nascent modern science of the seventeenth century, beginning with Bacon, depicts. Bacon and indeed Milton and Comenius attached great importance to the need for factual knowledge and technical training, as already said. Later in the century Locke, writing on education, said with some firmness that the technical treatises of the ancients, strictly useful if studied in preference to the literary and philosophical works, would produce only second rate citizens, presumably men of use rather than of erudition. Centuries before, Marcus Aurelius had denigrated banaustic skills in even more emphatic terms. These conflicting opinions have continued to be held, and will probably continue to be, but need not be discussed here.[19]

CHAPTER SIX

1700 to 1820

By the early years of the eighteenth century England was already beginning to achieve a position of leadership in advanced agriculture, a position that was consolidated in the second half of that century, and became an example worth copying throughout western Europe. This did not mean a complete break with the classical tradition, some elements of which in a dim sort of way contributed to encourage the new methods that were introduced. This is not surprising, but something of a commonplace since the writers of farming textbooks continued to pay lip service, some of them much more, to the revered authorities of antiquity. The foundation of this respect was, of course, the basic type of education received by the people who wrote the books.

Authors, to strike an obvious chord, must at least be able to write. Few undertake to write a book before they have read a good many other works. Consequently, most eighteenth-century aspirants for immortality, if only to the extent of producing a textbook on farming, the principal employment of their day, were men of education. Their readers were men of their own class. Indeed, it has been said with some show of accurate observation that 'until after the Civil Wars [in England] literature was the leisured accomplishment of amateurs . . . without a professional compulsion . . .' (though doubtless an inner urge) 'The reading public of Milton, Cowley, Waller, Dryden, Prior . . . and even to a degree thanks to the unique confluence of genius

and character of Pope himself . . . was probably roughly com-
mensurate with their social world as a whole.'[1] This continued
to be true through the times of the *Spectator* and the productions
of Addison, Steele, and others even to the reign of Johnson as
arbiter of a literary coterie. The same was probably (I have not
resolved this problem) true of France, Spain, Italy and Germany.
The intellectuals were a small proportion of the total population,
as always, even today; but in the early eighteenth century, the
scope of learning, the possible subjects, were less encyclopedic
and on the whole less technical than they are now. The foundation of learning was the Latin language and the Roman writers
first, and the Greek language and the Greek writers second. And
this was something that had continued through many centuries
with some interruptions when knowledge of these things became attenuated, only to recover as it did at the so-called Renaissance. This was the education that had been preserved by the
Church. No doubt the study of these languages and this literature
was excellent mental discipline, and, as David Ogg has put it,
obliged students to think out clearly what they intended to say.
On a different level there was, he claimed, a demand in many
parts of Europe for technical instruction and training in vernacular and foreign languages.[2]

It was to supply some part of this demand that farming textbooks had been produced since the end of the fifteenth century,
but it is undeniable that many of them owed a large debt to the
Scriptores rei rusticae. Others were more original, and, though
often very imaginative, contained some information derived
from practical experience. Still everywhere the classical writers
continued to be respected, often imitated, and were recommended
by some diehards like Nisard, who made his translations from the
Latin into French in 1844 with the intention that their instructions should be followed by the practising French farmer.[3]

There was more excuse for Richard Bradley, for the anonymous translator of Columella in 1745, and even for Adam Dickson. Richard Bradley, a man of great industry and somewhat
large pretensions, was of more capacity than was formerly
thought, he having been rather summarily dismissed as a mere
book-maker by some.[4] In 1725 he produced his *Survey of the
ancient husbandry and gardening . . . the whole rendered familiar to our*

climate; with a variety of new experiments. Bradley was not a modest
man. He dedicated the book to Townshend, either because of his
political status and influence, or because he was already interested
in farming. Bradley claimed that his book contained 'the sum of
the Greek and Roman Husbandry upon which . . . the Improvement of Land in England is chiefly founded'. In a sense this was
true because all human activities in the present owe something to
ancestral knowledge, although each generation must adapt it to
their own immediate needs supposing they do not accept it *in toto*.
Bradley thought it unfortunate that the husbandry of the ancients
had not hitherto been made familiar to English gardeners and
husbandmen, but admitted the difficulty of bringing Italian
management 'to be agreeable to our English climate'. Fundamentally he was right enough. The essential basis of growing
crops, the choice, preparation and enriching of the seedbed, are
elements of agricultural production, which are everywhere the
same allowing for differences in climate, elevation, slope, soil
consistency, drainage and so on. Bradley was sensible enough to
omit the auguries and astrology that were an integral part of
ancient wisdom. Two years later he published *The science of good
husbandry, or the Oeconomics of Xenophon showing the method of
ruling and ordering a family and of managing a farm to the best advantage* (1727).

Columella was translated by an anonymous writer and published in London in 1745 'with several illustrations from Pliny,
Cato, Varro, Palladius, and other antient and modern authors',
but the translator does not make it clear whether or not he intended this work for his contemporaries to be useful to English
farmers in their own then modern conditions. He was aware that
in Italy, France and Germany translations had been made into the
vernacular languages. Of these he seems to have been rather contemptuous. The intention of these translations was to make the
instructions they contained more ready to the hand of practising
farmers of the literate kind in the respective countries.

When in doubt about a reading this man followed that excellent edition published by the learned *Gesnerus*, Professor of Elo-
quence and Poetry at *Ottinghem* (Göttingen), who had done a lot
to restore the texts, not only of Columella, but all the *Scriptores
rei rusticae* to their original purity. Gesner is renowned. He issued

the *Scriptores* in Latin, both as a collection in 2 vol 4to in 1735, and later each author separately in small 8vo to form a collection. Vegetius and extracts from Martial are included in the quarto. Possibly this was more of an academic undertaking than to provide instruction for the farmer, the majority of whom would have been unable to read the books. As much or as little can be said of Peter Needham's edition of the *Geoponika*, published in Greek and Latin at Cambridge in 1704, an unlikely study for most husbandmen. It was reprinted in 1781.

It was for a more practical purpose that a collection of the *Scriptores* was issued in Italian at Venice between 1792 and 1800 under the general head *Rustici latini volgarizzati*, which included Cato, Columella, Pliny, Varro, Virgil, Palladius, Vegetius, and even Crescentius. Benedetto del Bene made a translation of Columella which was published at Verona in 1808, and an Italian version of Palladius in the same city, 1810. Del Bene's work was thought so well of that it was reprinted in 1850 in the *Biblioteca scelta di opera italione antiche e moderne* partly because the major part had appeared in scattered journals of his own day. His reputation was that of a golden writer about farming whose translations showed what a great agronome he was, not only in theory but in practice on his own estates. He died in 1825. It is not clear whether the translations were intended to be used as textbooks. The crop and fallow system of the ancients did not appeal to del Bene, who believed to some extent in the theories advanced by Camillo Tarello of Brescia in the late sixteenth century, ie, a four-course rotation including fodder crops, the system that had gradually spread all over western Europe from its origin in the Low Countries and its adoption in England whose example was generally admired. The area about Verona had been cultivated from time immemorial, or so del Bene declared, by means of irrigation through hydraulic wheels, the use of ample manure and frequent digging with the spade, but this was for market gardening.

Similarly translations from the *Scriptores* were made into French. G. Dumas made one with the title *Economique de Xenophon et de project de finance du même auteur* (1768). A six-volume edition with the title *de re rustica* was published at Paris in 1771 listed by Musset-Pathay as 'par Caton le Censor'. I have not seen this, but think it must have included other classical writers because Cato would hardly fill six volumes. Saboureux de la Bonneterie published a six-volume work, *Economie rurale par Caton, Varron, Columella, Palladius et Vegece*, in 1772. Can this be the same as the other? In 1765 an anonymous *Histoire de l'agriculture ancienne, extrait de l'Histoire Naturelle de Pline, Livre XVIII* attributed to Desplaces, with explanations and some remarks, came out in Paris. So late as 1824 Columella was translated into Castilian by de Sotomayer y Rubio under the title *Doce libros de agricultura que escribo in Latin . . . Columella*. Evidently the Spanish were proud of a succession of distinguished Spanish agricultural writers; Columella, who was born at Cadiz, the Spanish-Arab ibn al Awam, who lived at Seville in the twelfth century, and Alonso de Herrara in the sixteenth, not to speak here of lesser lights. M. C. Curtius made a German translation in 1769.[5]

It was with a very practical object that Adam Dickson, sometime minister at Dunse, Berwick, where farming was being improved, wrote his two-volume *Husbandry of the Ancients*, printed posthumously at Edinburgh in 1788. A quarter of a century before he had written a two-volume *Treatise on agriculture* (1762, 1770) because he thought the precepts in English books on farming unsuitable for Scottish conditions. He held a different opinion about the classical writers. Years before he wrote about them he had looked into some of the ancient rustic writers, and been 'agreeably surprised to find that, notwithstanding the difference in climate, the maxims of the ancient Roman farmers' were the same as those of the best modern farmers in Britain. This is rather an exaggerated view, but there are necessarily certain similarities. Dickson set out, with some measure of success, to collect under proper heads whatever was material to modern practice from the 'ancients'.

Though worthy of the consideration of the learned, agriculture had not, Dickson thought somewhat erroneously, occupied the attention of modern Europe for much more than a hundred years. He was convinced that something useful might be learned from the practice of the ancients, despite the danger of accepting 'old methods without knowing the reasons on which they were founded'. The book is indeed a careful analysis of the *Scriptores* with a contemporary commentary. Dickson was not only a student of the classical writings, but was also acquainted with Crescentius and an Italian he called Vincenzo Janara of Bologna, 1651, who was probably Vincenzo Tanara. No doubt he had read other writers' works that he does not specifically mention. It would be tedious, impossible here, to set out all his recommendations, but he does blame the British farmer for growing all crops on all kinds of soils whereas the Romans selected soils suited to the crops they intended to cultivate. And he makes great play with their cultivation of the artificial grasses, medica, lucern, sainfoin, and so on that were an important element in the current modifications of farming systems. In relation to these crops he could hardly fail to mention Jethro Tull and his theories.

Tull had a lot to say about Virgil's description of a plough, which had continued to be used in Greece, the eastern countries, Italy, and the Mediterranean littoral, and especially in Languedoc. Tull thought it strange 'that no Author should have written fully of the Fabric of Ploughs though many concerned themselves with the Stars and the bloody art of war'.[6] Even Tull had to refer to Virgil! Dickson thought that great poet had been misrepresented. No doubt. Is not everybody in some respect? But his fame was unceasing, and translations of his works multiplied as well as versions in his own language. The *Georgics* as well as the *Aeneid* had been favourite subjects of study through the centuries. Whether the *Georgics* can be accepted as a treatise on farming methods is quite another matter.[7] The judgement of the early eighteenth century is not altogether clear. The poems had been so often translated that new versions seem to the modern, but not apparently to contemporaries, to be a work of supererogation.

John Martyn, FRS, was one of those who aspired to fame by making a translation of the *Georgics* and indeed he is remembered for it. The first edition appeared in 1740–1, a fifth in 1827. Martyn was guilty of some exaggeration in the praise of his hero. Farming was a favourite subject 'of the most celebrated writers of antiquity', all of whom he finally summed up as inferior to Virgil who shone 'in a sphere far superior to the rest'. Later in his preface he made the somewhat equivocal remark that 'though the soil and climate are different from those of England; yet it has been found by experience, that most of his rules may be put in

143

practice, even here, to advantage'. To make these rules more valuable to English readers Martyn compared Virgil's advice with that of Aristotle, Cato, Varro, Pliny and Palladius, the last of whom 'wrote before the memory of Virgil's rules was lost in the barbarous ages'. By doing this he put an encyclopedic study of the classical writers into the hands of English readers who were without Latinity.

Martyn professed to have been most careful in his translation of Virgil's plant names, and this is a subject that has recently been returned to by an American scholar with some comments on the value of the fodder crops and their dispersion.[8] Martyn did not approve of Dryden's translation, a subject which had already been discussed by William Benson, but this was a literary controversy and not one related to the subject matter, farming. The book is a pattern of the attention to be given then and at a later date to literary exegesis, and not to the technical instruction provided by the classics.[9]

Otto von Münchhausen referred to Martyn's translation, and formed the intention of making a German version of the *Georgics* for use in schools to encourage the love of rural life and work. He placed Virgil in the forefront of all the Latin writers.[1] There were at least two other German texts published in the eighteenth century, ie, *Virgil's Lehrbuch von der Landwirtschaft* (Mulhausen 1792). The fourth edition of the other, made by Johann Heinrich Jacobi, was printed in Berlin, 1797. There were other translations, among them one into Spanish in 1793, but it is neither convenient nor necessary here to attempt a complete list of the various versions of this renowned and popular author.

The centuries-old practice of scholarly and literary exegesis continued to be exercised, and it is consequently often difficult to determine the precise purpose the authors of such works had in mind. Hesiod, like Virgil, frequently engaged the attention of students; but the number of reproductions and translations of his *Works and Days* approaches that of Virgil, and it is sufficient to mention them. Jacobi Vanierii of the Society of Jesus produced a poem, 'Praedium Rusticum', at the very end of the seventeenth century, ie, Paris, 1696. Written in verse it nevertheless follows the pattern of the encyclopedic works though not so lengthy, opening with the selection of a suitable site, continuing with all branches of farming, and completed with a kind of calendar of the year. He was apparently an apostle of Virgil but perhaps a trifle more practical.

Another work is a collection of extracts from various Greek and Latin writers, not only farming writers, arranged under heads, eg, praise of country life, description of agriculture, soil classification, crops and stock, etc, but whether it was intended to be of practical use is not clear. The author was Wolfgang Adolf Schronii, and the title *Syntagma de rebus rusticus et oeconomicus ex rei rusticae scriptoribus conscriptum* (Erfurt 1735). Adriano Kembter, a Premonstratensian canon, thought that rustic affairs were a great art and science, but neglected by the erudite. He therefore tried to distil the wisdom of the *Scriptores rei rusticae* in the course of discussion between three people, probably for the use of princes and eminent churchmen. The title of his book is *Veterum Scriptorum de re rustica praecepta in dialogus collecta* and it was printed at Augsburg. Finally, at least so far as here concerned, Gottlob Schneider edited *Scriptorum rei rusticae veterum Latinorum* in six volumes issued between 1794 and 1797 at Leipzig, but this was an attempt to purify the texts and to improve upon Gesner.

English farming writers, like those of other countries, continued to dazzle their readers in the light of the numerous authorities they cited, hoping perhaps that the lack of originality in what they advised might be concealed by this show of erudition. This habit of mind remained static with a good many authors though after the middle of the century there was a marked change. Another thing that these writers seem to have felt to be imperative was to supply, either in the preface or in the introduction to their theses, a brief, and occasionally not too brief, conspectus of the earlier literature of their subject often with references to the scriptural stories of the pastoral life, the annual bit of ploughing done by the emperor of China, the story of Cincinnatus, who left and returned to the plough, of the Roman patronymics that were the names of plants, and of the supposed witchcraft of Furius Criscinus. But some criticism of devout reliance on the classical farming writers was already being made in England early in the eighteenth century. It was an attitude of mind that was fairly general later in that century, and then not only in England. Throughout the period the classics were nevertheless reverentially referred to; but quite a large number of authors condemned their predecessors without mercy. Stephen Switzer, for instance, made no bones about his contempt for them. Most of the farming books he was acquainted with were translations from other languages (palpably an incorrect criticism for a man so widely read) 'and calculated for Soils and Regions quite different from ours'. Even Evelyn's *Terra* and *Sylva* were not free from this fault, the rules of *Sylva* being chiefly extracted from Pliny, Columella, etc, while Virgil's second *Georgic* had also been used. The practice of gardening had been conducted according to the rules laid down by the Latin writers ever since they wrote though they were 'mixed with too many superstitious Conceits'. Switzer had read the modern works as well as the classics.[11]

A slight indication of the trend of thought amongst English agricultural writers is John Mortimer's dedication of his book *The whole art of husbandry, being a full collection of what hath been writ either by ancient or modern authors* (1707) to the Royal Society, and his claim to describe new experiments. Like most of the other eighteenth-century writers of farming textbooks, Mortimer was an extremely well-read man, but he had been careful to adopt 'a plain Style for the Benefit of Vulgar Readers, the Culture

Fig 11 *Woman reaps, man sleeps, from an old print of probably the seventeenth century*

of Lands being left almost entirely to their Management'. This was not entirely true even in his own day, and became less so as time passed. He was convinced of the accuracy of the classical system of crop and fallow if restricted to comparatively infertile soils in England, and he praised Virgil's advice on burning stubble, which continues even today in East Anglia if not elsewhere.

The Rev Walter Harte was another who complained that 'agriculture is now [1764] the drudgery of the lowest part of mankind, and not the amusement of the brightest and most elegant geniuses' as it had been in the great days of yore. Indeed, 'agriculture would soon carry another aspect in this Kingdom if every gentleman were a true rural economist according to the sense of the antient writers on husbandry'. He was convinced that there was no modern knowledge of the technique of farming equal to that of the classical writers, more especially Virgil, but including all the others both Greek and Latin. Here again was a writer who had read all the earlier literature, ancient and modern, including the *Geoponika* and Crescentius. He compares Columella and Cowley rather daringly because the latter cannot be considered a technical writer.[12]

Much the same opening remarks continued to be made in English, and many European writers on agricultural textbooks throughout the eighteenth century. It would be tedious to compare them all in detail. A brief list will be found in the chapter notes.[13] English writers were so conformist in the preparation of their prefaces and introductions, but there was a growing tendency to look to field work (perhaps developing into experiment)[14] rather than to accept uncritically the advice of earlier writers or the technical traditions of the 'vulgar' farmers. This tendency had begun to develop, amongst scientists rather than agriculturists in the seventeenth century, and appears clearly in the Abbé de Vallemont's work, *Curiositez de la nature et de l'art sur la vegetation*, first published in 1703 at Paris, and translated by Bishop William Fleetwood into English in 1707. De Vallemont was as well read as his compeers. He knew of van Helmont's experiment, and was acquainted with the work of other seventeenth-century scientists. His main interests were the relation between soils and vegetation and in the control of plant diseases by seed steeping. His book was translated into German as well as into English. It was an indication of the change from consulting the authority of the ancients, though this continued for comparative reasons, to the authority of practice supported by trying out new theories in the field. The lectures given by Sir Humphrey Davy to the Board of Agriculture was an intermediate culmination of this approach. 'It was part of a wide effort to convert farming from a mere art of blind processes into a rational system of science.' André Bourde called it 'a work of a new conception'.[15]

Jethro Tull was perhaps the most striking of the proponents of the new outlook. His theories of plant nutrition, mistaken though they may have been, and the fame that resulted first from his experimental work and later, in the 1730s, from the publication of his books, were an immense stimulant to novel thinking both for and against his theories.[16]

Opinion was divided about Tull both in his own day and later, the fate of most innovators, but his work and his books had a profound effect on the development of farming all over western Europe. Robert Brown of Markle acknowledged in 1811 that agriculture was of great antiquity. He thought it had been taught by Noah to his sons, but assumed that materials were wanting 'from which a progressive history of the art can be composed'. He nevertheless made some attempt to do this in an elementary way similar to other agricultural writers. He did not think much of Tull whom he estimated to be 'an inventive genius who possessed more imagination than solid sense to direct it'.[17] This was an opinion not shared by David Henry, who declared that 'the only theory of husbandry that has hitherto received the sanction of modern approbation is that of the celebrated Jethro Tull . . .' Henry seems to have broken away from tradition completely. He does not appear to refer to classical authority at all though some of his advice is the same, necessarily so, because the technique of farming, ie, preparation of the seed bed, choice of soil, manuring, seeding rates, etc, had remained static for at least 2,000 years.[18]

Tull himself said he was persuaded to write and describe his theories by the publication of Martyn's translation of Virgil, but this is probably only part of the story. Other pressure had been put upon him to set out his experiments and their results, and to record his methods and newly designed implements for the benefit of the agricultural world at large. This he did, as every history book now relates, in the third decade of the eighteenth century. Far more attention was drawn to his ideas and machines by the commentary made by du Hamel du Monceau, the experiments of Châteauvieux, and the translations of du Hamel's work: Spanish in 1751, Dutch in 1762, German and English in 1764,[19] as well as the international exchange of some of the controversial writing anout these articles and processes. These led to the observation of results in the field rather than obtaining advice from the pundits of classical times: though their books were still read, for a diversity of reasons, and some of their advice followed because it was good for all time, or so it seems in the twentieth century.

André Bourde has recently remarked upon the change in outlook in France after 1750 when a different attitude, critical instead of complete acceptance, was adopted. This reconsideration was not confined to France. It was general over western Europe, and particularly in England where, as is well known, not only Tull's theories and field work, but also the increasing adoption of roots and artificial grasses were amply demonstrating that slavish following in father's footsteps was a mistake. This was more pronounced in England, whose influence was certainly already being felt on the continent.

Langethal has expressed the same opinion about Germany as Bourde did about France. The authority of the classics was sinking. Men had long since realised that the German climate and conditions required very different methods from those of the old Romans, and had come to distrust the ancient literature. By 1750 farming had become fashionable, and a school of empirical writers had begun to flourish. This new approach to the problems of farming technique became more emphatic after the end of the Seven Years War (1770).[20]

The principles of plant nutrition had engaged the attention of scientists in the seventeenth century. Possibly the controversy about Tullian theory stimulated speculation, if not experiment, in the mid-eighteenth century. Francis Home's *Principles of agriculture and vegetation* was published at Edinburgh in 1757. It was translated into French in 1761 and German in 1763. Almost concurrently Wallerius directed a dissertation by Count Gustavus Adolphus Gyllenborg, which was printed in Latin and Swedish at Uppsala in 1761 under the title *Agriculturae fundamenta chemica*. This book was translated into German by Krünitz, Berlin, 1764, into French in 1766, and into English by John Mills in 1770, into most other European languages, and into Latin in France in 1791.

This exchange of scholarship was no new thing. It had been usual for centuries, learning having pretty well always transcended national boundaries. The process had been facilitated by the printing press; but the exchange of agricultural information by translation of national (or native) works is a point that will be returned to. It has awakened a good deal of interest of late, as indeed it may have done in the past.[21]

The Italian writers were roundly condemned by the Rev Walter Harte for their slavish adherence to the precepts of the classical writers, for whom he himself was not without a measure of appreciation. 'Such of these Italian writers on husbandry as did not concern themselves with translations made the ancients of their country their text and model', he wrote when describing sixteenth- and seventeenth-century works. This criticism continued largely true in the following hundred years although the Italians could not possibly have entirely escaped the changes of opinion that were taking place then. Besides these changes of opinion there were the practical consequences of the introduction of new crops from overseas, as Benedetto del Bene pointed out in 1808 when he translated Columella. The classic writers did not know such things as rice, maize, or sorghum, the potato, some oil and dye plants, the mulberry tree and the silkworm.[22]

Important as these things were, especially maize and the potato, for human aliment, and the industrial crops for manufacture, it was perhaps less on them than on the so-called artificial grasses and root crops that supplied better nutrition for the livestock, that improved farming was based. Some of these leguminous crops, lucern, clover, sainfoin, were known to and recommended by the classical writers as the copious quotations made by the Rev Harte amply demonstrated to his readers. Protagonists of these crops had been writing in vernacular languages since the sixteenth century. They were grown in some places then, eg,

Page 151 (above) *A happy scene, from Robert Brown of Hill Farm, Somerset, in* The Compleat Farmer (1759); (below) *a ploughman using the Kentish turnwrist plough, from John Boys,* General View of the Agriculture of the county of Kent (1796)

Flanders, Piedmont and southern France. By the end of the seventeenth century a few English farmers scattered over the countryside had adopted them. Tull was amongst these, though sainfoin with which he experimented was grown in Berkshire and elsewhere along the chalk escarpment in his day. These fodder crops and Tull's theories were the basis of the breakaway from the standard farming process of hundreds of years—as is well known; so is the chance that the new cropping system came to be called the Norfolk four-course rotation. It is almost invidious to mention such well-known events, but as the foundation of modern development they have to be.

Already by 1753 it could be said by an Italian writer that a new system of farming had been shown by Tull through the translation made by du Hamel. The subject was expanded some years later by Francesco Griselini, who acknowledged the talents of Tull and du Hamel, but claimed that the seed drill had been given its final touches in Italy, a claim based on national pride rather than on fact. However, Griselini admitted that the English were the first to study agriculture with good results, most other nations following their example—a large claim with regard to Italian farming which still resembled that of the Romans in many places at a much later date. A type of seed dropper fixed between the plough stilts is illustrated in his book. By this time, too, the work of Francis Home was known in Italy, having apparently been translated into Tuscan and printed in Milan in 1763. Griselini also wrote a pamphlet exhorting the parish priests to educate and instruct the peasantry in the best methods of farming. Naturally the seed drill continued to occupy the authors; many were invented in Italy as elsewhere during the succeeding decades, and reference was made to the uncertain history of earlier attempts to design such a machine. Similarly the Norfolk system of crop rotation and manuring was discussed, notably by Antonio Campini.[23]

To discuss Italy is rather to put the cart before the horse because it was via France and Switzerland that the new ideas were transmitted from England to other European countries, just as the new ideas had reached England from Brabant and Flanders. *The influence of England on the French agronomes, 1750–1789* is a subject that has been exhaustively treated by André Bourde

(Cambridge 1953). It will therefore only be dealt with briefly and simply in this place, but fully enough to demonstrate, as Bourde does, that there was a more or less conscious break with tradition about 1750.

This phenomenon was not limited to agriculture. As every schoolboy is aware the year 1760 has been registered as the date of the birth of the so-called Industrial Revolution. All now realise that no exact date can be given to this romantic parturition, but engineers of the eighteenth century did learn 'that industrialisation implied a complete break with the past'. It put a period to the age of craftsmen, something that is deeply regretted by the nostalgic modern as it was by the Morris school in the last century.[24]

A somewhat similar change took place in the world of literature. In the seventeenth and early eighteenth centuries most writers, poets and essayists subscribed to the classical tradition, and produced works modelled on these admired authors to interest men, and sometimes women, of their own class with similar education. But it was not only the starving pamphleteers of 'Grub Street' who began to make a change. The accepted and acceptable authors were also making cracks in this formidable nexus. Pope in his *Discourse on pastoral poetry* had subscribed, naturally enough, to the classical tradition, but his friend Gay looked back, not to the supposedly happy shepherd and gambolling lamb of the Greek hills where nymphs sported, but to the much more realistic memory of his life in Devonshire. His poems described rural life as he had seen it, not in the laces and fripperies of a court fashion. The renowned James Thomson often did the same in his 'Seasons', though his poem is marred by his floating off into the empyrean of India and other recondite places. Nevertheless, he was a pattern that many imitated including such lower-class aspirants to literary fame as Stephen Duck, Dodsley, and others. John Dyer was completely realistic in his *The Fleece* and it has been described as a textbook. However, he paid sufficient attention to the prevailing mode to introduce the idea that ancient kings and chiefs were shepherds, as undoubtedly our remote nomadic forefathers may have been. The process of being instructive came to its apogee when Wordsworth in his 1815 preface declared that this was the underlying principle of

154

poetry 'as the poem of Lucretius, the *Georgics* of Virgil, *The Fleece* of Dyer, Mason's *English Garden*'.²⁵

This development in England was quite dissimilar to that in France where literature seems to have been wholly an apanage of the court, as the plays of Corneille, Moliere, etc, even including the works of Voltaire, of Rousseau with his noble savage, and so on, indicate. There was, however, a real attention to nature paid by a great man, Buffon, and a lesser light, de la Pluche in his *Spectacle de la Nature* and there were technical tracts and books, many heavily indebted to English writers and English husbandmen. The most influential of these was undoubtedly du Hamel du Monceau's *Traité de la Culture des Terres* which appeared in six weighty volumes between 1750 and 1756. This work created a great furore all over Europe and was widely translated. Besides an exegesis on Tullian philosophy and a description of his implements as modified by du Hamel, the book comprised records of a very large number of experiments made by admirers in scattered areas all over France, and on the estate of Château-vieux in Switzerland.

Its general distribution was perhaps as much a part of the then prevailing 'francomania' as it was of the similar 'agromania' in France against which Desplaces uttered a protest in 1762, a protest about as effective as Canute's demonstration of his inability to control the tides, although he made the pertinent remark that if everybody had remained shepherds or ploughmen there would have been no need for books on farming. Nevertheless, he had read most of them, but held tenaciously to the common opinion that agriculture exacted the most severe labour from the farmer, the result of all his labours being nothing more than poverty. He did not say precisely that he agreed, but he did remark that Patullo thought most of the books useless, only to be perused by amateurs. Desplaces knew Francis Home in the French translation *Les principes de l'agriculture et de vegetation* (Paris 1761) and discussed recent developments in the theory of plant nutrition as well as local practices in different parts of France, which he thought a much safer guide to follow than the books, both ancient and modern.²⁶

Maybe he was right, but the recorded word is the basis of change (?). At any rate it is a good many years since Marc Bloch noticed that the agricultural revolution was part of the movement of thought in the Century of Enlightenment, which implied a rationalisation of practice coupled with a contempt for tradition. The new husbandry about which there was so much controversy could mostly be included in the developing experimental science, which, he thought, owed so much to England. Although no one would deny that eighteenth-century Europe followed or tried to follow the lead of French cosmopolitanism and French thinking, thus creating a French Europe from Moscow and St Petersburg to the shores of the Atlantic, this was a pattern only followed by the wealthy and aristocratic. It did not impinge upon the rural population, the peasantry who were poverty-stricken and overworked.[27]

The year 1750 can be accepted as a point of departure for a new enthusiasm for farming amongst people of more comfortable social status than the peasantry, that is, the nobility, rich and poor, and possibly some successful bourgeoisie who were intent on becoming members of that class; but there was obviously some interest amongst the home-dwelling, as well as the court nobility before that date. There was even some literature for them to read. The most massive and maybe the most important contribution to these didactic works was Louis Liger's *La nouvelle maison rustique ou economie générale de tous les biens de campagne* which first appeared in 1702, and like its predecessors Estienne and Liebault, and Olivier de Serres, went through a good many editions, each of which seems to have been subjected to some revision. This is a claim that is made in the preface to the third edition, 1721. The book was reprinted at Amsterdam in 1750. It follows the plan of the encyclopedias of instructions for living in the country that became customary after Crescentius and, in a more concise form, Varro: choice of site, building, all branches of farming and gardening, livestock in health and illness, even to cooking, and is thus in the proper line of descent from the classics. Some of the new crops from overseas are treated. Another encyclopedic work was Noel Chomel's *Dictionnaire Oeconomique* (Paris 1709). Bourde thinks that it played an important part in the elaboration of agricultural doctrine, through its author's vast but uncritical erudition. Chomel cited all the accepted sources and discussed English agriculture and English writers.

His compilation was translated into English from the second edition of 1718 by Richard Bradley in 1725, and into Dutch as *Algemeen huishoudelijk*, published in parts between 1778 and 1793.[28]

Just before the crucial date there was some slight tendency towards a development of science, in other fields as well as agriculture, particularly natural science, which is of importance in relation to farming. It is perhaps impossible to repeat too often that the study of nature, fundamental to farming, which is after all almost absurdly biological, by scientists, literati, poets and artists, had been brought to birth in the Italian renaissance. It was mainly in the interests of luxury and learning but came to bear rather heavily on the ideas of eighteenth-century people whose interest in farming was completely practical. They wanted to eat! And indeed to drink, but not particularly the pure lymph! The plan of the *Encyclopaedé* (1748–65) underlines the change in outlook. Diderot, or at least some of his collaborators, believed in science rather than in tradition, and hoped to stimulate further progress in understanding the world of nature and consequently the principles which were the fundamental basis of farming practice.[29] The example of Norfolk in England was accepted by Diderot and other Frenchmen as it was later in Italy by Campini and others as the guiding light of the practical farmer. Diderot's *Lettre écrite du Comté du Norfolk* was an example of the French acceptance of this guidance, and the *Lettre* was copied by others who used the same title around about 1750. Bourde thinks that this was partly due to the snob value of Townshend's title in the painfully class-conscious French court. The lead in reclamation and light land farming given in north-west Norfolk was practical, and closely followed in different places in Europe. It was, for example, the spur to the activities of such men as the Marquis de Turbilly though Bourde suggests that he got his ideas from Germany. Be that as it may he began to reclaim his estate in 1737. He published his *Mémoire sur les défrichements* in 1760 (Paris, 4to), but when Arthur Young visited the place in September 1788 it had once more declined to its original rather primitive state. The practical results of the Norfolk example were not repeated as a consequence of the Tullian philosophy, which did not lead to the adoption of his principles generally either in England or in

Europe. It had, however, the important effect of turning some landowners and others, not excluding the literary enthusiasts, away from tradition to the examination of what experiments in the field could teach them.[30] This in turn led to controversy and discussion, a wholesome clearing of the air likely to uncover new knowledge—as indeed it did.

Its application was limited for various reasons in spite of the work of the government. Even in the literature there continued to be some esteem for the classics right down to the mid-nineteenth century. André Bourde's massive three-volume work has dealt most comprehensively with the extensive agricultural literature, and it is not only impossible to do so here, but would be a work of supererogation. A few examples will indicate the trend of thought in the particular aspect under consideration. They have not been selected for any other reason beyond their accessibility in England.

For example, Angran de Rueneuve in his *Observations sur l'agriculture et le jardinage* 2 vol (Paris 1712) used the conventional approach to his subject by reference to scripture, Homer and the Latins with a glance at the *Geoponika*. The authority of Virgil and Columella was sought by Dr Wolfius. An English version of his *A discovery of the true cause of the wonderful multiplication of corn* was printed at London in 1734. He professed to have made some experiments on how to stimulate tillering and prevent blast. In this connection he referred to de Vallemont, Dodart, etc. Dr M. Kulbel (physician to the king of Poland) wrote a *Dissertation sur la cause de la fertilité des terres* in 1741, which was published at Bordeaux. He subscribed to the classical idea that soil fertility could be judged by its natural ecology. One Caraccioli published *L'agriculture simplifiée d'après les regles des anciens* in 1749 (according to Bourde, *Agronomes*, I, p 447), which was apparently a return to classical authority on the very eve of change. Desplaces, author of *Preservatif contre l'agronomie* issued his *Histoire de l'agriculture ancienne extrait de l'histoire naturelle de Pline, livre XVIII* in 1765, and declared the vaunted novelties 'n'est autre chose que nos méthodes connus'. He referred with pride to the works of Bernard Palissy, paysan de Xaintonge, to Buffon and others. The philosophical implications of farming were commented upon by M. Arcere, *De l'état de l'agriculture chez les*

Romains (Paris 1777), who supported his conclusions by a vast bibliography of classical authors, some of whom are only writers on rural life and the pleasures of rusticity, and not agricultural textbooks. Rumours of classical theory are also to be found in M. Barthes' *Mémoire d'agriculture et de mechanique* (Paris 1763), especially in methods of soil classification (p 113–4). Colour, taste, smell and touch were criteria, but the best was touch. If soil stuck to the fingers, it was rich: but Barthes did not subscribe to the classical test of soil structure by digging a hole and filling it up again. This process was not very reliable because in different weather conditions the soil was of varying compactness, perhaps the first criticism of the Roman method. Barthes believed (p 1) that the four elements of the ancients completed the growth of plants, but in very different ways according to the diversity of soils.

One pronounced anglophile (?) was M. Patullo, whose *Essai sur l'amelioration des terres* came out in Paris in 1758. The book was intended to describe all he had seen in Scotland and England. He praised du Hamel and Châteauvieux, but entered a caveat to the effect that if Tull's method had not been followed in England there was some doubt of its efficacy in France. The rest of the book follows more or less the usual pattern describing then modern farming. J. Bertrand wrote an essay on water meadows in 1764 which depends to some extent on English practice, and a fairly simple *Elémens d'agriculture fondés sur les faits et les raissonnements à l'usage du peuple de la campagne* (Berne 1775). Both works won prizes from the Société Economique de Berne.

The result of all this writing and discussion was to make some people think with two minds. For instance, Ebaudy de Fresne, *Traité d'agriculture* 2 parts (Paris 1788), was strongly in favour of his countrymen following the example of England in reconstruction of its farming methods, but nevertheless stated that the Romans had remarkable success which had only been possible as a result of the perfection of their cultivation, a somewhat dubious assertion. The French should therefore take them for their models, adopt their methods, do blindly as they did, and thus obtain the same results. No less elevated a person than the Abbé Rozier, who organised an encyclopedic work, *Nouveau cours complet d'agriculture* in six volumes (Paris 1809), assured his

readers (p 139) that this was already the state of European farming . . . 'in France, England, Germany, everywhere the pattern of
Roman agriculture was preserved in spite of the numerous
revolutions these countries have endured since the fall of the
Roman Empire'. He was, however, moved to describe current
practice in each country (p 141ff). Musset-Pathay underlined
this by remarking that most of the systems described in the
Geoponika were still followed in 1810 (Bourde, *Agronomes*, I,
p 451).

Acquaintance with the novel principles of English farming was
spread amongst the French upper classes, not only by the observations of individuals who had visited England, but also by translations of English farming textbooks into the French language.
Amongst such translations are *L'agriculture complète de Mortimer*
in 1765, the English version having been published in two
volumes in 1707. The translation of Francis Home's *Principles of
agriculture and vegetation* has already been mentioned. Henry
Home, Lord Kames's book *The gentleman farmer* (1776) became
Le gentilhomme cultivateur ou cours complet d'agriculture which is a
rather exaggerated description of its pretensions. Young's *Political
arithmetic* and his *Travels in France* in three volumes were obvious.
Varlo, as *Nouveau systême d'agriculture* in three volumes, reached
a fourth edition in 1775. Cuthbert Clarke's *True theory and practice
of agriculture* was translated in 1779. A version of John Philipps's
attractive poem, 'Pomona ou le cidre', appeared in the early
years of the century. Forsyth's book on fruit trees was translated
in 1803, and Tatham's on irrigation in 1805. Finally, a version of
William Marshall's work, translated and rearranged in five
volumes, was issued in Paris as *Maison rustique anglais ou voyage
agronomique en Angleterre* in 1806. This list makes no pretension
to being complete. It is an indication of what was happening.

Oddly enough one M. Paris, an architect, thought it worth
while to translate Dickson's *Husbandry of the ancients* as *L'agriculture des anciens*, the two volumes being published in 1801-2.
Charles Pictet, the Swiss agricultural writer, translated Emanuel
Fellemburg's work from German as *Vues relatif à l'agriculture
Suisse et aux moyens de la perfectionner* (Geneva 1808). Fellemburg,
who had established an experimental and teaching establishment
at Hofwyl, thought well of the Norfolk system, but thought that

too much had been made of the Tullian precepts by du Hamel and Châteauvieux. Consequently the system had become completely discredited. The drill required simplification. Other machines were urgently needed, harvesting machines, threshing drums, etc. Obviously Fellemburg was looking to the future. He makes no play with the classics.

The current interest in Merino sheep led to the translation of Colonel Alstrom's book from the Swedish as *Essai sur la race des brebis à laine fine* printed at Metz in 1773.

Besides translations of English books into French there were some from the German, but the Germans owe a much greater debt to the English and French than vice versa. One of the most important German works to be translated was that of Johann Beckmann, *Grundsätze der teutschen Landwirtschaft* (Göttingen 1769) of which a second edition was printed in 1790. The French version was entitled *Élémens d'agriculture* and was made by M. Sylvestre in 1791. This work is valuable for the present purpose because it contains a bibliography that pretends to include all the farming textbooks of western Europe from classical times—though it is far from complete. Another useful feature is a list of all the agricultural societies with which the author was acquainted.

The Germans were always interested in compiling lists of authorities from the days of Camerarius to modern times, but in one branch of history they were in advance of the rest of Europe, ie, in the writing of agricultural history. By the beginning of the nineteenth century there were already some studies in the history of the plough, eg, Andreas Berch, *Methodus investigandi origines gentium ope instrumentum rurali* (1795); Ginzrot, *Die Wagen und Fahrwerke der Greichen und Römer* (1817), which included a section on the plough; Fr G. Schultze, *De aratri Romani forma und compositione* (1820); Lastereis, *Sammlung om Maschinen, Instrumentum und Gerätschaften* (1821-3). A Frenchman, Mongez, had entered these lists with his *Mémoire sur les instruments d'agriculture des anciens* (1815). K. H. Rau completed this series, in a sense, with his *Geschichte des Pfluges* (1845). There are later works on the subject—a great many of them—but in parallel some German scholars were stimulated to write histories of their country's farming. Perhaps the first of these was Karl Gottlieb Anton, *Geschichte der teutschen Landwirtschaft von den ältesten Zeiten bis zum*

Ende des funfzehnten Jahrhundert, 3 vol (Gorlitz 1799–1802). Next was C. Fraas, *Geschichte der Landwirtschaft, 1750–1840* (Prague 1852), a Gekrönte Preisschrift. It was followed by Chas Ed Langethal, whose *Geschichte der teutschen Landwirtschaft* was published in Jena in 1854. There have been successive studies of the subject ever since. Max Guntz published his *Handbuch der landwirtschaftlichen Litteratur* in 1897, an invaluable source book, but by no means so complete a catalogue of agricultural works in all languages as it somewhat pretentiously claims.

For some reason interest in this subject did not develop in England until the end of the nineteenth century although it has expanded largely in the last few decades. A Frenchman, Rougier de la Bergerie, wrote a *Histoire de l'agriculture française* that was published in 1815; an Italian, Gabriella Rosa, a *Storia dell'agricoltura nella civilla* in 1883. There are many later works in all these countries.

Perhaps the reason for the early interest of German scholars in the development of their national farming systems was partly the result of the disasters of the Thirty Years War and the slow recovery that followed it, on which all these historians agree. There had been a beginning to the long series of *Hausväterlitteratur* before the war, and the number of writers of these compilations increased in the eighteenth century. These books were much of a pattern, following Crescentius, and the scheme of the encyclopedic works of the sixteenth century. It may be bold to suggest that von Hohberg's work of late seventeenth-century date, was the outstanding example of this kind of production. It was reprinted five times, the last time in 1749, which was about the evening of that sort of book. His work is that of a highly educated man, and his teachings owe a great deal to the classical writings. There can be little doubt that his example was followed by the writers who succeeded him in this genre.

After about the mid-eighteenth century the textbooks were written to a rather different pattern. There was less originality in their titles. The earlier books were not very numerous. Abel found fourteen examples in Göttingen University Library dated between 1670 and 1750. Like von Hohberg, whom they copied and expanded, these books called upon past wisdom as well as each other. They made little attempt to break away from tradition.

One thing they did in the midst of all their encyclopedic pre-occupations with rural living was to compile lists of the precedent literature as the foundation of their work. It is by no means a necessary consequence that the compilers had read all the books in their lists, however heavily underscored the implication may be. Von Hohberg's work is a vast folio, one of those works that strikes awe to the heart: his successors expanded the scheme into three-, four- or five-volume productions. Some were repeatedly reprinted.[31]

The first break with tradition in Germany was not made as a consequence of the new approach to agricultural technique made by Tull, and so enthusiastically followed by du Hamel and Châteauvieux. It was made by the Kameralists, who began to write books to guide their princely employers—as well presumably as each other. Their writings were doubtless inspired by the tuition provided by the two chairs of Kameralistik established in 1727, one at Halle, the other at Frankfurt am Oder. Perhaps this new science would be called estate management today—in the sense of the management of landed property. The prototype was published by Simon Peter Gassers of Halle in 1729 with a lengthy title reduced here, *Einleitung zu den Oeconomischen, Politischen und Cameralwissenschaften*. Scholarly respect for previous writers did not, of course, vanish in a flash. That is impossible. For example, Dithmars, who occupied the chair at Frankfurt am Oder, writing a couple of years later supplied a classified bibliography including Xenophon, the Latin authors, the *Geoponika* down to current productions. A similar long list of authorities is to be found in Zincke, *Allgemein ökonomischen Lexikon* (1743), which is largely occupied with agricultural problems. Abel thinks von Justi the most important German Kameralist of the second half of the eighteenth century, a judgement it is impossible to oppose. There were, of course, other writers on this aspect of agrarian affairs dealing not only with estate management, but also with practical farming.[32]

The need for experiment in the field was not emphasised by the Kameralists who were mainly concerned with legal and financial problems. It was only in the 1750s that, as in other countries, farming became fashionable in the higher social circles, and there was a new, or rather modified system of thought, a change in the general mental atmosphere that was a symptom of the times. Many widely held beliefs, superstitions if you like, were being called into question all over Europe. Langethal indeed rejoiced that the useless references to the classical writers were abandoned because the theorists had finally realised, in the 1750s, that the German climate and other circumstances which dictate the form agriculture shall take were quite different from those of the old Romans; something Fraas had remarked before him. Consequently a group of writers came to the fore who were the founders of a rational agriculture—as Thaer later called it.[33]

These men were much the same class as the Kameralists, but had a rather different though allied approach to the problems of the day. Doubtless this new thinking owed something to what is now called the mental climate of the time. They founded what is known as the empirical school, which may mean that they looked at what was being done in the fields, and possibly tested new ideas on what would now be called demonstration plots.

The earliest of them was Christian Reichart, who published his *Land und Gartenschatz* in six volumes between 1753 and 1755. Langethal, however, thought that von Rohr, who added natural science to his agriculture (IV, p 250) had initiated the new form of agricultural literature (IV, p 287). Nevertheless, it is true that Reichart's work was the beginning of the new approach. He was antagonistic to the current respect for old writers, but had some ideas derived from Ellis and Tull as well as du Hamel, and tried to write some sort of history of agricultural practice, but this was, in fact, no more than praise of rural pursuits. Johann Gottlieb Eckhart's *Vollständige Experimental ökonomie über das vegetabil- animalische* appeared in 1754. To some extent this follows the accepted pattern of opening with Adam's fall and a very rough outline, with little real context, of history, but this is really only in support of his subject. Another man, who is accepted by some as a joint founder of the new approach, was Hofrat Hagedorn, whose *Landwirtschaftlicher Haushalter* appeared in 1755. A more definite supporter was Johann Georg Leopoldt with his *Nützliche und auf Erfahrung gegrundete Einleitung in der Landwirtschaft* (1759). Leopoldt claimed that his book contained only what he had himself experienced. He subscribed to the theory that was becoming accepted all over Europe that the foundation and keystone of all

164

economy is fodder. He believed that a substantial increment of fodder could be obtained from carefully constructed water meadows, and was a protagonist of the four-course rotation, including cabbages and peas in the fallow break. Clover and roots had been cultivated in the sixteenth century, as Heresbach pointed out, but that was in a limited area, and these crops had not spread through Germany by the mid-eighteenth century, when Schubart, der Graf von Kleefeld, gained his honorific title for his propaganda. Even though these crops and the new importations from America, potatoes, maize and tobacco, were cultivated, the area on which they were grown was not yet extensive.

One result of this variety of systems was the publication of descriptions of local farming. Otto von Münchhausen, for example, in his *Der Hausvater*, 3 vol (1765), supplied a description of current Westphalian farming in his first volume, ie, the three-field system. C. F. Rosenow, described Mecklenberg 'köppelwirtschaft' in his *Versuch einer Abhandlung von Ackerbau und Köppelwirtschaft* (1762). C. W. Schumacher did the same in the following year with his *Gerechte Verhaltnisseder Viehzucht zum Ackerbau aus der verbesserten Mecklenburgischen Wirtschafts Verfassung*. Münchhausen mentioned much earlier local works, on Saxony and Brandenburg in 1730; Silesia in 1712.

Münchhausen was another man who attempted a comprehensive bibliography of agricultural literature from the earliest times. Johann Beckmann, *Grundsätze der teutschen Landwirtschaft* (1790), did the same. It was a task to defeat the most assiduous and enquiring scholar as the justly renowned Max Guntz found at the end of the last century.[34] Beckmann was convinced that a sufficient history of farming had not yet appeared in any country, a doubt that may perhaps be echoed even today.

The works cited here are not all the German agricultural books published in the eighteenth century, but are sufficient to show the change from reference to authoritative literature: to an examination of the facts of farming with some lingering backward glances at what may be called the standard literature with its legends of kings and consuls who farmed, including the annual performance of the Chinese emperor. In Germany this change was more marked because the theorists emphasised the differences between their country and Italy. There were marked variations in the local practice of farmers in the divergent physical conditions of their environment, a circumstance which made comparisons possible and instructive. This was not an isolated phenomenon. The farmers in the north and south of France, and some in less distant localities, were obliged to suit their operations to their situation. The same thing is now known to have been true of eighteenth-century England.[35]

Both French and English books were translated into German in the eighteenth century. Some Germans travelled in France, but their interest seems, like the literature, to have been mainly in horticulture, fish breeding and hippology until the nineteenth century.[36] The French were, indeed, themselves learning from England both by exploratory travels, by translations of English books, and from original writers. A larger number of English works was translated into German. Nearly fifty titles are appended to Müller's paper in the *Agricultural History Review* already cited. The books range in time from John Mortimer in the early years of the century to Arthur Young and William Marshall at the end. A complete set of the *Annals of Agriculture* is included. Frederick II, too, not only employed Christopher Brown, but sent some young men to England to report upon the methods employed there.[37]

Many societies were formed in Germany for promoting farming, supported by the nobility and landed gentry in the latter part of the eighteenth century. Schools of agriculture were founded and the literature continued to increase. A definitive position was reached when Albrecht Thaer began his experimental work and teaching, and began to write. His *Einleitung zur Kenntnis der englischen Landwirtschaft* came out in three volumes in 1804. This book not only described the variety of English practice, but offered criticisms which, if accepted, would make these systems appropriate to Germany. Of course this was not all Thaer did, but this is not the place to discuss his other activities. The break with the classical tradition seems to have been complete by the turn of the century; in Germany and other countries, the result was supported by the development of chemical science, particularly those sections of the subject dealing with plant life.[38] This assertion takes account of the fact that Johann Gottlieb

Schneider produced the *Scriptorum rei rusticae veterum Latinorum interpreted* . . . at Leipzig in four volumes.

South of the Alps the Italians continued to work, at least in a literary sense, upon the classical heritage, and to some extent on the Greek ancestors of their own culture. This strain runs through their farming textbooks of which a large number of one kind and another continued to be written during the eighteenth and early nineteenth centuries. It is very doubtful whether the peasants and mezzadri ever came into direct contact with any of these. Some of the teachings, novel or traditional, may have reached them through the example of landed proprietors or their agents, where, as in Piedmont, everybody was more or less affected by the new fervour of agricultural studies, which in the second half of the eighteenth century, as Giuseppe Prato put it so long ago, constituted in the whole of Europe one of the most characteristic manifestations.

Despite this new fervour, crop yields in Piedmont were still very low in the middle of the century and not much better when Arthur Young 'toured' the district at its end. Wheat, rye, white and red maize, oats and other cereals were cultivated, but wheat only gave a return of $5\frac{1}{2}:1$; often only $3\frac{1}{2}:1$; rye only $5:1$; down to $3:1$. Arthur Young was scornful of yields then only $6:1$, occasionally $5:1$. The hay crop averaged $26\cdot56$ quintals per hectare.[39]

In a rough way the agricultural literature of eighteenth-century Italy can be divided into three kinds. The first is the translations of the classical writings; the second is reprints of earlier works, with or without commentary; the third is contemporary vernacular writing. Nothing like a complete bibliography of these works will be attempted here partly because only a limited number is to be seen in this country; partly because the sample, accidentally composed, demonstrates the degree of dependence of the then modern writers on their predecessors, ancient and modern.

The translation work of Benedetto del Bene in the last decade of the eighteenth century has been mentioned. There were other assiduous translators. Virgil, naturally enough, continued to be popular, and at least two versions were put out in the eighteenth century, that of Castelvetro Francesco Contuti, at Modena in

1757; that of A. Antonion Ambrogi in 1774. There may have been others. Bernadino Corradi published his *Versione Italiana del decimo libro di L. G. Moderato Columella* . . . in 1754; Brazuolo Paolo Milizia *Esiodo. Le Opere e i Giorni tradotto*, in Padua in 1765. Another erudite author was Carolus Aquinas. He had studied the *Scriptores* and issued his *Nomenclator Agriculturae* from Rome in 1736.

Reprinting of old Italian books was not extensive, but it did happen. Alemanni's poem 'Coltivazione' was one example. Three issues were put out during the eighteenth century, the first in Padua in 1718; the second in Verona in 1745 and the third in Lucca in 1785. In addition Vincenzo Benini of Padua wrote *Notes on Alemanni with the twelve books of Crescentius added* in 1745. Rucellai, whose book on bees was often printed with Alemanni, was republished in 1746. Clemente Africo came out again in 1714. Vincenzo Tanara was reprinted at Venice three times, in 1700, 1731 and 1745: Camillo Tarello at the same city in 1772 and Gallo's *Vinti giornate* at Brescia in 1775.

Only a small number of original farming and agricultural economic works was written in Italian in the eighteenth century, and many of these are not to be seen in England. However, a sufficient proportion can be examined to enable some idea of the general tenor to be obtained. All the authors respect antiquity very much, and there are many references to the honour in which farming was held by statesmen, soldiers and others. The ritual ploughing of eastern potentates is another common theme. The current pattern of the work being done only by the rude unlettered peasant of the horny hands and stooped back is as generally deplored.

The furore about farming that had arisen after about 1750 makes it little matter for surprise that the majority of the works published appeared in the second half of the century. An exception, which is, as the title makes clear, little more than a collection of precepts from earlier pundits, is the anonymous *Ricordi di Agricoltura raccolti da migliori Autori di coltivazione antiche e moderni*, the third edition of which came out in 1735 at Florence. It was intended for use in Tuscany.

Cosimo Trinci wrote three books, the first with the resounding title *L'agricoltura sperimentato ovvero regolo generali sopra l'agricoltura* (Lucca 1726). It was reprinted at Milan in 1851 in the collection *Biblioteca scelta di opera italiane antiche e moderne* (vol 563). Trinci made frequent reference to the classical writers, especially Virgil, but also supplied a calendar of operations for the guidance of his contemporaries. According to the Rothamsted Library Catalogue the book was reprinted seven times in expanded form by 1805. Copies of all these editions are filed there. This library also possesses his other works, *Nuova trattato d'agricoltura* (Venice 1778), and *Raccolta d'opusculi appartenenti all'agricoltura* (Venice 1768), the character of which may perhaps be gauged from its title. This writer seems to have had a long and busy life.

But the new science and practice was beginning to make its impact on the Italian theorists and writers by the mid-century. In spite of frequent references to Columella, Varro, Palladius and Theophrastus as well as Holy Writ, Antonio Campini knew the writings of Francis Home; and Wallerius was aware of Tull's theories and well acquainted with the principles of the Norfolk husbandry so that he was able to write a description of it. He also appreciated the value of the novel crops, the theories of Tarello and other early Italian and many French writers. He thought Tull's drill difficult to make, and very heavy for use in Italy. Griselini's ideas were more appropriate to his native soil.

G. F. M. Cacherano wanted to expand the area of cultivated land, and set out his ideas under the title *De Mezzi pur introdurre…la coltivazione e la popolazione nell'agro Romano* (Rome 1785). This countryside had, he said, been deserted by the peasants in Roman times. They went to Rome to get free rations. Constant reference is made to the classical writers in support of his contention that country life and living is best for one and all. The necessary equipment for each man is estimated. The reclaimed land was to be set out in townships surrounded by the inhabitants' holdings, laid out in squares in the old Roman fashion, and not in spheres of influence in concentric circles as laid down by von Thunen in *Der isolierte Staat*. Simple tools, the hoe and the spade were to be used as well as the plough and the harrow, something that was said again in a wider context by Fabroni in his *Istruzione elementari di agricoltura* in the following year. These things were probably of the same pattern as those used long before by Roman farmers.

Most other hand tools have retained their general design since prehistoric times. The ploughs used in the Bolognese plain were indeed precisely those described by Virgil. By adding two ears to this simple construction, as Virgil remarked, a sort of moulding plough was made. If provided with one ear only this made a primitive mouldboard, and when held at an angle could in effect turn a kind of furrow, perhaps not a complete inversion of the soil. Filippo Re in 1798 wished and recommended the abandonment of the *piota*, plough with two ears (Virgil's *duas aures*) by the peasants of Emilia, but it continued to be used in Bologna and other places well into the nineteenth century. The mouldboard plough was more usual in Tuscany. A wheel plough with double mouldboards (ears?) was used in Savoy, where, according to M. le Marqués Costa in his *Essai sur l'amelioration de l'agriculture dans les pays monteux . . . Savoie, Chambery* (1774), farming was quite barbarous, but the peasants would not accept novelties though they prepared the land for some crops with an *araire*. The Piedmont *araire* had the two ears required by good sense, but fixed ears on the Savoyard plough was against nature. Four oxen were employed to haul the Savoyard wheel plough, but only two in Piedmont though the implement was equally heavy. This book is contemporary, but M. le Marqués does refer to Camillo Tarello in support of some of his more modern ideas.[40] The controversy on this subject, however, seems to have been more literary than practical. Bandini held the definite opinion, readily acceptable, that eighteenth-century Italian farming was little different, technically, from that practised in Roman times, although new plants had been introduced by the Arabs and from America. Only in Lombardy about 1730 was the system of continuous cultivation adopted with the abolition of fallow and the introduction of fodder crops, the major change in technique since Roman times. The forage crops, trefoil, lupines and lucern were, however, known to the Romans and were occasionally taken as a catch crop by them. Campini greeted them as the plants of progress, which indeed they were all over western Europe.[41]

Both Ferdinando Paoletti and Giuseppe Nenci admitted their dependence upon old authors, but equally both had some glimpses of the new ideas. Paoletti in his *Pensieri sopra l'agricoltura*, 2 vol, 2nd ed (Florence 1789), thought the practice of farming was not so good as it was in Columella's time. Two things were essential to improvement, the first to free the peasants, the second to increase the number of beasts on Tuscan farms. This writer was a priest, and thought priests should teach the peasants better farming methods. He cited Albertus Magnus as well as the classics and the Italian pundits who had preceded him. Nenci in his *Reflessi sopra le piu frequenti e necessari operazione della coltivazione* (Florence 1791), like a lot of others, wrote a good deal about olives and vines, as well as the cereals. His precepts are more or less modern for that day in spite of his respect for the old authors.

Books on particular subjects were not infrequent. The familiar crops, the vine and the olive, were of great importance, and deservedly received attention. Hemp was an economic product. The innovations such as tobacco, maize, rape and potatoes were naturally discussed at some length. Diseases of plants and animals caused devastation. Bianchi wrote on the cattle epidemic that decimated the livestock all over Europe, besides Italy, in the first half of the century. Francesco Ginanni struck a new note as well as an old one in his *Della malattie del grano in erba. Tratto storica-fisica* of 1759. There are copious references to the Greek and Latin writers, and the index contains a list of books in which reference is made to plant diseases throughout the ages. *La meteorologica applicata all'agricoltura* was treated by D. Giuseppe Toaldo, in a new edition of his work dated 1786 at Venice. This man had numerous connections all over Europe, and was widely read.

Few of the eighteenth-century Italian writers dispensed with some reference to the ancient literature or relied upon the unsupported authority of the systems they recommended, or failed to quote precedent vernacular works. This is the more comprehensible because they were calling upon their (in some sort) ancestors, and the methods of the ancients. Inevitably there was some break with tradition, and the Napoleonic Wars caused marked social changes, but not changes in the pattern of farming. In many parts of the country it continued in the good old-fashioned way well into the nineteenth century. This is not to say, of course, that there were not some districts where the new (but so old) ideas were put into practice with marked advantages to the innovators as in other countries.

Spain did not escape the prevailing influence of the day. At least twenty books on farming were published in that country during the eighteenth century.[42] It has not been possible to examine these in detail on account of an incomplete knowledge of the language. Some of these were translations into Spanish from other tongues. The majority came out in the second half of the century, but Louis Liger was printed in Spanish in 1720 with the title *Economia general de la Casa de Campo Obra muy util de Agricultura*, and was said by the translator, Dr Francisco de la Torre, to be considerably augmented. References are made to Herrara so that there was necessarily some dependence on classical authority. Du Hamel was translated twice, once in 1773 on trees, and in 1751 on agriculture. This modern note was repeated in a translation of G. A. Gyllenborg (Wallerius's essay) by Casimera Gomez de Ortega as *Elementos naturales y quimicos de agricultura* in 1794. Abanos translated Richard Kirwan on manures in 1798.

Another book is said by Ramirez to have been published first in French, next in English by John Nicholls, an author who cannot be identified, and finally translated into Castilian in 1775 as *Dos discoursos solve el Gobierno de los Granos y Cutivo de las Tierras*. Ibn al Awam was translated from Arabic into Spanish by Josef Antonio Banqueri as *Libro de agricultura* (Madrid 1802). There was a translation of Daubenton in 1798, and in that year Francisco Luis Laporta issued his *Historia de la agricultura España*. There was a number of quasi-original works, some of an encyclopedic nature, and some dealing with a particular subject. Spanish horses and mules were famous and profitable, and deserved special treatment in books about them and veterinary science: one a translation of Bourgelat. There were some books dealing with the problems of particular areas, eg, Estremadura and Galicia with the Asturias, but it would be fatuous to pretend to a precise reading of these, and possibly more so merely to repeat Ramirez's list, because this would not bring out any dependence upon earlier writers, and more especially the *Scriptores rei rusticae*. Incidentally, the first Spanish agricultural society was founded in 1768: others were set up between 1780 and 1810. Members of these societies and other grandees formed the public for these books, and were doubtless full of interest in the Tull-Duhamel theories and suggestions. The so-called artificial grasses no doubt occupied space in most of them, but so far as can at present be judged it was not until 1797 that a work devoted exclusively to this subject was published. It was *Prados artificiales* by José Manuel Fernandez Valligo.

The people of the Netherlands, at least the farmers, seem to have been more practical than literary, unlike their peers in other countries. The vernacular literature on farming produced there during the eighteenth century was very scanty, but for the benefit of the richer classes who lived in the country or had houses or estates there, a fairly large volume of horticultural texts was published. Their peculiar circumstances had led them to make advances in farming technique in the late Middle Ages that were adopted in England in the seventeenth and in other countries in the eighteenth century, and then only on a limited and local acreage.

Dr van der Poel has gone so far as to say that before the eighteenth century there was no original agricultural literature in the Netherlands. Most agricultural books were compiled from the Roman writers or were translations from foreign authors.[43] One early eighteenth-century book of this kind was drawn from a German work by Georg Andreas Agricola entitled *Neu und nie erhörter doch in der Natur . . . universel Vermahrung aller Baume . . .* 1716–7, of which another edition was published in 1772. It came out in the Netherlands in 1724 as *Nieuwe en ongehoorde dog in de Natur welgegronde . . . boomen*, etc. Richard Bradley translated it into English in 1721, as he did Noel Chomel's *Dictionnaire Oeconomique* in 1725, the original French having appeared in 1718. This, too, was turned into Dutch as *Algemeen huishoudelijk . . . woordenboek* by J. A. Chalmot in 1748: a second edition appeared in 1778–93. Du Hamel was apparently the axis on which change turned in the Netherlands as elsewhere on the continent of Europe. Van der Poel says (*op cit*) that men theorised and without exact knowledge about du Hamel's trials, and a few large landowners began to make similar experiments themselves. He adds that van Engelen translated du Hamel into Dutch, including a commentary and some Dutch experiments. The new husbandry therefore came to the Netherlands via France and not from England. C. van Engelen's book was *De nieuwe Wijze van Landbouwen*, and was printed in four volumes (Amsterdam 1762–5). Both Chomel and van Engelen emphasised the importance of the new row crop culture: they did not need to impress upon the farmers of the Netherlands the value of fodder crops, nor of obliterating the fallow break. Both these works are enormous, and only a great gift of patience would enable anyone to read them through. In some sort they gave advice that was based on practical experience, possibly mainly in other countries, but advice that was independent of the authority of the classics.

Quite a few books were written about the epidemic diseases that ravished flocks and herds all over Europe during the eighteenth century, but these relate to measures of control, of treatment, and of the political circumstances which developed as a result of the disasters. It is not relevant to discuss them here.

CHAPTER SEVEN

Summary

I<small>T</small> is becoming almost painfully commonplace to say that history, as seen by the historian, depends upon his personal taste, and that this in turn depends upon the social environment in which he lived or lives.[1] The evidence used is invariably selected either by choice, something extremely difficult to control, or because it is all that remains. Worse still, many of the documents used were not prepared with an eye upon the interests of posterity, but for immediate practical purposes, and are of value to historians only in so far as they can be interpreted to produce a meaning which the document was not intended to supply. None of these disadvantages can apply to textbooks. At different dates agricultural textbooks were written to guide farmers in their current work. They can therefore be read and understood as if the modern reader were seeking for the same information. The only obstacle is that the number of agricultural textbooks written from classical times to the present day is so vast that selection is forced upon the enquirer, and it may be that the selection is not the best that could be made for several reasons. The most important is that only those books that can be seen can be read, and then only if the reader possesses a knowledge of the language in which a book is written. Libraries, too, do not contain a complete set of all these works, even those in the country where the book was written. Consequently there is a further restriction upon what can be studied. The balance is large, and presents all the usual problems of time and space.

Some doubt has indeed been expressed by various persons, not without pretensions to be heard—Thackeray for one—whether accurate history can be written at all. Setting up the muse of history to be knocked down, Thackeray wrote, 'You bid me listen to a general's oration to his soldiers. Nonsense! He no more made it than Turpin made his dying speech at Newgate . . .' A lesser, insignificant light, one John H. Steggall reported that some are ready to insist that all history is fabulous, and that the very best is but a probable tale, artfully contrived and plausibly told.[2]

This sort of criticism can hardly be made of what is in the end only a comparison of the contents of books about the same subject written at different times and in different countries. No claim can be made to any originality of approach. None is possible. Perhaps it may be plausible to say that besides being in a restricted way a bibliography, incomplete it is true, the result will also be history. Almost two hundred years ago G. Gregoire laid it down that the only way to write the history of agricultural science is to write the literary history, an undeniable but incomplete asseveration. He substantiated his idea by producing a work that is largely a bibliography of European farming books. In spite of this he said that the number of writers who deserve to be cited as classics is very small, the rest being farmers of the indoors and compilers, an opinion that even then was not very original. He claimed that Columella and Varro were the equals of the most celebrated of contemporary writers on the subject,[3] an opinion that was presumably shared by Nisard, who translated the *Scriptores rei rusticae* into French for the guidance of his fellow countrymen so late as 1844. This was somewhat a work of supererogation despite the continuance of classical methods on the Mediterranean coast. It was not a new idea. Seventeenth-century men, as no less an erudite than André Bourde has pronounced, did not miss the shortage of vernacular textbooks, being satisfied with copious editions of Greek and Latin writers. Not only did they find the methods satisfactory for current use, but by reason of the minimal rate of the evolution of technique and science, works even more than a century old were not useless.

Bourde does not think the difference between Virgil's knowledge of mechanics, physics and chemistry and that of de Serres was marked, the established ideas having prevailed for some

176

1,600 years. The changes made in the eighteenth century were not the organic development of the traditional principles, but inspired by the external influence of those parts of the country where farming methods were clearly different from Virgil's precepts. But the ancients could not be wholly neglected. They were basic education apart from farming, something that the writers of farming textbooks were no less anxious to demonstrate than other people. The books were consequently overcharged with references designed to make a parade of learning. Even the Physiocrats were not free from this fault in Bourde's opinion.[4]

These writers had the best of excuses for their rather uncritical acceptance of and admiration for the classics. Not only was the literature of Greece first, and that of Rome somewhat second, accepted as the final example of beauty in writing unlikely to be surpassed, but also there was an opinion that these books contained, if not the last word in technical wisdom, at least a stimulant to further thought. The Greek and Latin agrarian writers, it has been said, from Xenophon to Virgil, demonstrate a widening realisation of man's ability to change the natural order,[5] something that is a necessity as soon as arable farming is undertaken. This was implicit rather than explicit in the rules of the *Scriptores rei rusticae* but following their advice would have made it clear, even if every practising farmer had not already known it. Whenever a weed is removed or, on a more extensive scale, a seedbed for a crop selected by a farmer is made, the natural order is changed. But knowing this full well did not lead the ordinary farmer to experiment with different methods. When his family needs were cared for with a trifle of odds and ends left over for sale to provide requirements that could not be made at home, there was no reason to work any harder or to aim at a large net product so avoiding what is now called full employment and reserving ample leisure.[6] Perhaps this is one reason for the numerous Church holy days of medieval times when no work was done and some sort of festivity was organised in most localities. There is, of course, the other possibility that no higher yields could be secured when using primitive tools and implements, almost unselected seed, scanty supplies of manure, and suffering the other handicaps of a pre-scientific age.

The continuance of long-established practice in the field as well as in the books is undeniable. At the end of the eighteenth century the ancient system of taking two corn crops after a fallow, said to be familiar to the Roman husbandman and practised in Virgil's time, was followed, the few cattle being fed on straw with hay, only a trifle, as a supplement.[7] Amongst the few cattle the plough team of oxen must have been numbered. Both oxen and horses were used for this job as contemporary pictures of country life show, and the controversy which raged for centuries about the relative virtues of these two sorts of animal traction. The ox team must still have been a familiar sight in the early eighteenth century as Pope's pastorals indicate. In his 'Autumn' he wrote:

> 'While lab'ring oxen, spent with toil and heat
> In their loose traces from the field retreat;'

In his 'Spring':

> 'Sing then and Damon shall attend the strain
> While yon slow oxen turn the furrowed plain.'

Even here is some harking back to the classical pastoral, Damon not being an English name. But Pope was writing for his compeers, many if not most of them the landed proprietors who ruled not only England but nearly all western Europe. Their education and outlook was based on classical learning, and therefore in some measure what used to be called hide-bound, although some of them were aware of the new ideas that had been stimulated a good while before by such philosophers as Francis Bacon, Lord Verulam. The English Royal Society was, of course, composed of such men. Besides this governing class of landed aristocracy there was a growing aristocracy of money made in trade, banking and so on, more particularly in the Netherlands, Italy, West Germany, not to speak of England and France.[8] People of these occupations are frequently open to new ideas, and since wealthy men in those times usually sought to buy land in order to become members of the accepted ruling class they may have been more susceptible to the fresh outlook because they would undoubtedly have tried to make their estates pay, something that has been said a good many times before.

The change in the approach to science, technology and

literature that had been occurring for some centuries can be said
to have reached a stage in the eighteenth century that made it the
basis of the modern world. The idea of experiment was not new
when Roger Bacon expounded it in the thirteenth century, nor
was the accurate observation of nature such as was made by
Albertus Magnus. Both were emphasised by men of a later date
like Bacon and William Harvey, who wished to found their
knowledge on observation and experiments made by themselves
or their contemporaries. The agricultural writer, Walter Blith,
made the first suggestion for an experiment in growing a succession of crops.[9]

Fig 12   *A dairy scene, sixteenth century, from Mattioli, 1598*

It would be inappropriate to indulge in extensive literary
history in this place, but some reference must be made to the
differences between the poets and essayists of the eighteenth century and those who had graced the world before them. These
people were influenced by and influenced the temper of their
times, and reflected the modified approach to their subjects,
especially those that were bucolic, to be found in what would
now be called other disciplines. This is partly the reason for the
reference to Pope above.

Pope indeed has a tincture of the artificiality of the classical pastoral, of which he was not unaware. Gay, his contemporary and friend, despised the flavour of Arcady, and refused to write of nymphs and dryads, or shepherds piping on oaten reeds, a performance which may have been possible, but is incredible if oat straw was used to make the pipe. Contemporary shepherds slept under the hedge, not under myrtles, nor had they to fear the attack of wolves because there were none, at least in England. The village maidens followed routine tasks in the cowshed and dairy, helped in tying up the sheaves, and driving the pigs home to their sties. There was no gaucherie in this realistic approach to country or pastoral themes. It is found also in John Philipp's *Cyder* and in Dyer's *Fleece*, parts of both being little more than metrical textbooks, but of a less doggerel kind than Tusser's pedestrian verses. Not to labour the point, describing country life as it was actually lived became a usual preoccupation of the eighteenth-century poet, who adorned the fantasies of a non-existent golden age to which earlier writers had looked back. Some elements of realism practical farm jobs, employed Thomson, in his re-nowned *Seasons* and it has been said that he was influenced by agricultural writers like Richard Bradley and Stephen Hales. Crabbe, Bloomfield and Clare, Cowper, Wordsworth, all wrote of what they had seen around them. Goldsmith and many another have touched upon everyday and contemporary happenings. Some were to introduce science into their lucubrations. Erasmus Darwin was one of these.[10] But the influence of Virgil on nearly all bucolic verse can scarcely be denied, especially his realism.

Quite apart from the derivative nature of rustic verse there was every reason why eighteenth-century poets should laud country life and labour, not as a pleasing poetic fancy disguised in the imagined personalities of the Greek pantheon, but as a record of what people were actually doing. Not only was the world until the latter years of that century necessarily mainly agricultural for reasons of food supply and of the slow development of technology, but because of the organisation of society, ruled as it was by the landowners. Willing as they may have been to increase their revenues, these people were richer in land than in money, or so it may be supposed. Supplies of money (capital) required to pay the costs of improvement and the materials necessary were, it is said, scanty so that agriculture was stimulated more than industry, which demanded much more expensive stock, buildings and equipment.[11]

In these circumstances the farming community, at least the literate part of it, was prepared to listen to the preaching of the prophets of change. And the changes made were severely practical, aimed at the growing market of an expanding population. Some part of the literature was written by scientific men, mainly naturalists interested in investigating the problems of plant life, its structure and nutrition. The exact knowledge these investigators were acquiring should have been of immediate consequence to rural society, but it did not make its full impact until the early decades of the nineteenth century. This new learning, a very different matter from the humanities of the Renaissance, was the basis of technical change. Science had to be appreciated and used by men doing the work to which it could be adapted.[12]

This was not a new phenomenon. Many useful things had been made by ingenious men in and before the Middle Ages. Who, for example, made the first plough all those millennia ago? Later the horse collar made traction easier, the stirrup gave the horseman a firmer seat. The clock began to regulate the hours more exactly than the church bells: the mariner's compass enabled sailors to sail the seas with more certainty of making the landfall they hoped for. How did these inventions come about? The goad that causes a man to devote himself to an invention is one with many prongs, and cannot easily be made clear. 'Necessity', the proverb runs, 'is the mother of invention'. Forbes is inclined to the opinion that the remarkable expansion of overseas trade in the seventeenth and eighteenth centuries invited invention. Maybe. He was on more certain ground when he said that technical training owed little to the universities, and that farmers have always tried to reduce the work of the farm, both their own and that of their employees by using gadgets in place of men.[13] One result of this was that the new engineers were not brought up in the classical tradition any more than were the guild apprentices of an earlier time.

The new mathematics were not likely to have a great effect on farming, but their use in the manufacture of precision measuring instruments may have done something to stimulate the interest that urged several people to try to adumbrate a mathematical theory of the plough, and ploughing techniques. These efforts developed what was known as the Rotherham plough, originally imported from the Low Countries, but improved by the work of various plough mechanics as time passed. There is no evidence to support the theory that this branch of knowledge was of help in designing the seed drill and the threshing drum, but a guess that this was so may be made.[14]

The introduction of the new fodder crops, roots and artificial grasses, has been sufficiently discussed, not only in the preceding pages but by almost every writer about farming history in all the west European countries, so that no more than a passing mention need be made here. Some of these crops, more particularly the legumes, and perhaps the roots in Gaul, were known to the Romans and possibly abandoned because of the uncertainty of life and the scarcity of cattle to eat them, and travelled by devious routes for nearly 2,000 years before being widely cultivated in the northern countries of Europe. The new introductions from overseas were of importance. Because these were unknown to the ancients their cultivation formed a real innovation.

Emphasis has always been placed upon the shortage of manure, which confronted farmers until the scientists came to their help. Most of the *Scriptores rei rusticae* make great play with the importance of conserving all kinds of domestic, animal and vegetable refuse to enlarge the manure heap. Marl was used in Gaul, possibly in Britain, and soil mixing was advised by Columella. Almost every writer of a farming textbook in almost every language is preoccupied with the problem of fertility, of supplying the crop with the nutrition it required to flourish and yield heavily; but it was not until science came to the help of the farmer that this problem was more or less solved.[15] It is to be feared that he did not immediately have much contact with the subject, and Dr R. A. C. Parker has said that even so great a man as Coke of Norfolk was supremely contemptuous of these chemists.

The spread of literacy in the lay world doubtless had a great deal to do with the increase of productivity, yields of crops, improvement of livestock in every way, and so on, because the farmers could consider and appreciate the advice they were being offered. Until the sixteenth century this ability was confined to a single class, a monopolistic property of monks and clerics. Both peasants and lay aristocracy were illiterate because they had no reason and little opportunity to be otherwise. Education became more widespread, but not, of course, anything like general after the invention of printing. It was not an absolute necessity for the rural population, who learned their work by doing it as their parents directed. But it would be improper to trespass within the boundaries of the history of education here. Suffice it to say that literacy increased, that the study of the classics was the prime concern of most schools and universities. There was also in the past half a dozen centuries a proportion of the educated people who looked to their surroundings rather than to the cult of humanism which developed at the Renaissance, and was later to some degree suspended by the scientific curiosity of what may be termed the modern centuries. The final result was that the authority of the classics as literature continues to be respected even in this iconoclastic age, but that the authority of the *Scriptores rei rusticae* as farming textbooks has been superseded.

Notes

Chapter One

1 Rostovtzeff, M. *Social and economic history of the Hellenistic world* (Oxford 1941, II. p 1181–83).
  Hauger, Alphons. *Zum romischen Landwirtschaft und Haustierzucht* (Hannover 1921, Preface).

2 Derry, T. K. & Williams, Trevor I. *A short history of technology to AD1900* (Oxford 1960, p 55).
  Xenophon. *Treatise of Householde*, tr Gentian Hervet (1499–1584), in *Certain antient tracts concerning the management of landed property reprinted* (1767 p 22).

3 Glotz, Gustave. *Ancient Greece at work* (1926 p 34).
  Seignobos, Charles. *The rise of European civilisation* tr C. A. Philip (1939 p 38).

4 Honigsheim, Paul. 'Max Weber as historian of agriculture and rural life', *Agricultural History* (XXIII. July 1949 p 193).
  Toutain, Jules. *Economic life of the ancient world* (1930 p 34–6).

5 Hesiod. *Works and days* tr Hugh Evelyn White (Leob Classical Library 1954 *passim*).
  Baumeister, A. *Denkmaler des klassichen Altertumer* (Munich 1884, p 10–12).
  Mahaffy, J. *A survey of Greek civilisation* (1897 p 59–62, 115).
  Waltz, Pierre. *Hésiode et son poème moral* (Bordeaux 1906, *passim*).
  Seignobos. op cit p 30.
  Burn, Andrew Robert. *The world of Hesiod* (1936 *passim*).

6 Michell, H. *The economics of ancient Greece* (Cambridge 1940, p 20–1).
  Fussell, G. E. 'Population and wheat production in the 18th century', *History Teachers Miscellany* (May 1929 p 66–7).
  Nicholson, Edward, *Men and measures. A history of weights* (1912 p 34).
  Wilson, A. Stephen. *A bushel of wheat* (Edinburgh 1883, p 28–9).
  Clark, Colin & Haswell, M. R. *The economics of subsistence agriculture* (1964 p 77–80).
  Brentano, Lujo. *Das Wirtschaftsleben der antiken Welt* (Hildesheim 1961, p 32ff as to the necessity for importation of grain to supply Greece after the first century BC).
  Piggott, Stuart. *Ancient Europe from the beginnings of agriculture to classical*

antiquity (Edinburgh 1965, p 47, 252; p 94 confirms the yield and consumption stated).

7 Tod, Marcus N. 'Economic background of the 5th century BC' (*Cambridge Ancient History* V. 1927 p 13–14).
Toutain. p 33.

8 Jacks, L. V. *Xenophon, soldier of fortune* (1930 p 217, 221).

9 Xenophon. *passim.*
Nisard, M. *Les agronomes latins* (Paris 1844, p 49 fn, cf Theophrastus).

10 Semple, Ellen Churchill. *Geography of the Mediterranean region; its relation to ancient history* (New York 1931, p 301).
Toutain. p 107.

11 *Idem.* p 37.
Glotz. p 39–41, 250–7.

12 Baumeister. p 12.
Cary, M. 'The general condition of Greece in 386 BC' (*Cambridge Ancient History* VI. 1927 p 57).
Rostovtzeff. II. p 1180, 1186–92.
Heitland, W. E. *Agricola: a study of agriculture and rustic life in the Graeco-Roman world* (Cambridge 1921 p 16–25).

13 Paul-Louis. *Ancient Rome at work* (1927 p 51–5).
Rostovtzeff, M. *Social and economic history of the Roman Empire*, hereafter cited as '*Roman Empire*' (Oxford 1926, p 13–14, 275).
Heitland. p 203.
Toutain. p 174–7, 190, 191, 207.
Hauger. preface and p 8.

14 Frank, Tenney. *Economic history of Rome to the end of the Republic* (Baltimore 1920, p 46).
Paul-Louis. p 163.
Anon. 'On the agriculture of the Romans' (*Quarterly Journal of Agriculture* II. 1831, p 346).
Heitland. p 179.

15 Cato. *De agri cultura* tr W. D. Hooper & H. B. Ash (Loeb Classical Library 1959, I. i, vi–ix).

16 The distinction Ernest Brehaut tr *Cato the Censor on Farming* (New York 1933, p xxiii) saw here is difficult to substantiate. cf Nisard, cited in note 9, and Cato, p 2.

17 Cato. XXIII–LIV.

18 Mommsen, Theodor. *History of Rome* (1862 II. p 362–71).
Rostovtzeff. *Roman Empire* (p 9–10, 13–21, 23).
Heitland. p 14, 165–6.

Paul-Louis. p 156–9.
Savoy, Émile. *Hammurabi à la fin de l'Empire Romain. L'agriculture à travers les âges* vol II (Paris 1935, p 68).

19 Frank, Tenney. p 1–12, 53–8.
Anon. *Quarterly Journal of Agriculture* II (1831 p 10–1).

20 Rostovtzeff. *Roman Empire* (p 30–1, 61–5).

21 Fussell, G. E. 'Marl, an ancient manure' (*Nature*, Jan 24 1959).

22 Schröder-Lembke, Gertrud. 'Romische Dreifelderwirtschaft'. *Zeitschrift für Agrargeschichte und Agrarsoziologie* (XI. April 1963, p 25–33, and works therein cited).
White, Kenneth D. 'The efficiency of Roman farming under the Empire' (*Agricultural History* XXX. April 1956, p 87).
Lizerand, Georges. *Le régime rural de l'ancienne France* (Paris 1942, p 8–9).
Behlen, H. *Der Pflug und das Pflugen bei den Römern und in Mittel Europa in vorgeschichtlicher Zeit* (Dillenberg 1904, p 31–3).
Stevens, C. E. 'Agriculture and rural life in the later Roman Empire' (*Cambridge Economic History* I. 1941, p 89, 92–3).
Frank, Tenney. p 98.
Bath, B. H. Slicher van, *Agrarian history of western Europe* tr Olive Ordish (1963 p 54–63 esp p 59).

23 Martialis. *Epigrammatum* (III. lviii).
Boissier, Gaston. *The country of Horace and Virgil* tr D. Havelock Fisher (1896 p 11, citation of Saint Beuve. *Étude sur Virgile*, p 32).
Hadzits, George Dupue. *Lucretius and his influence* (1935, esp on Virgil, p 31–3).
Dill, Samuel. *Roman society from Nero to Marcus Aurelius* (New York 1957, p 174–80, 196).

24 Thorndike, Lynn. *Medieval Europe* (1920 p 35).
Heitland. p 210–11, 246, 310, 360.
Rostovtzeff. *Roman Empire* (p 90–4, 192).
Paul-Louis. p 267–73.
Lot, Ferdinand. *La Gaule* (Paris 1947, p 80, 113–15).

25 Hubert, Henri. *The greatness and decline of the Celts* (1934 p 85).
Chilver, G. E. F. *Cisalpine Gaul: social and economic history from 49 BC to the death of Trajan* (Oxford 1941, p 129–45).

26 Caesar. *Gallic war* tr H. J. Edwards (Loeb Classical Library. 1909, p 347, 355).
Tacitus. *Germania* tr Maurice Hulton. *Idem* (p 271, 285, 287, 301).
Hoek, J. J. Spahr van der. *Geschiedenis van de Friese landbouw* (1952, I. p 37–41).
Roupnel, Gaston. *Histoire de la campagne française* (Paris 1932, p 102–3).

Wührer, Karl. *Beiträge zur ältesten Agrargeschichte des Germanischen Nordens* (Jena 1935, p 5, 11).

27 Honigsheim. p 206.
Abel, Wilhelm. *Geschichte der deutschen Landwirtschaft von frühen Mittelalter bis zum 19 Jahrhundert* (Stuttgart 1962, p 17).
Lizerand. p 26–7.

28 Below, Georg von. *Geschichte der deutschen Landwirtschaft des Mittelalters in ihren Grundzugen* (Jena 1937, p 18, 26).
Dopsch, Alphons. *Economic and social foundations of European civilisation* (1953 p 5, 21, 39).
Latouche, Robert. *The birth of western economy: economic aspects of the Dark Ages* tr E. M. Wilkinson (1961 p 30–6).

29 Paul-Louis. p 62, 321.

30 Wright, Thomas. *The Celt, the Roman and the Saxon* (1852 p 205).

31 Ausonius. *Poems* tr H. G. Evelyn White (Loeb Classical Library. 2 vols).
Sidonius, Apollinaris. *Poems and Letters* tr W. B. Anderson. *Idem.* 2 vols.
Lizerand. p 13–15, 17.
Moss, H. St L. B. *The birth of the Middle Ages* (Oxford 1935, p 26, 29, 245–6).
Roger, M. *L'enseignement des lettres classiques d'Ausone à Alcuin* (it may be mentioned that none of the rural writers are included in the index).
Lot. p 365–6.
Sutherland, C. H. V. *The Romans in Spain 217 BC to AD 117* (1939 p 100–2).
White. p 88 (largely speculative).
Hayward, Richard Mansfield. *The myth of Rome's fall* (1958 p 10, 13, 103–4, 115, 137, 170, 172).
Seignobos, Charles. *History of the French people* (1939 p 29, 30, 43, 49).

32 Hawkes, Christopher. 'The Roman villa and the heavy plough' (*Antiquity* IX. 1935, p 339–41).
Collingwood, R. G. & Myres, J. N. L. *Roman Britain and the English settlements*, 2nd ed (Oxford 1937, p 208–21).
Stevens. p 103.
Rivet, A. F. I. *Town and country in Roman Britain* (1958 p 100–1, 117–25).
Clarke, R. Rainbird. *East Anglia* (1960 p 122, 125–8, 131–2, 137, 150–1, 163).
Applebaum, Shimon. 'Agriculture in Roman Britain' (*Agricultural History Review* VI. 1958, p 66–86).

Chapter Two

1 Power, Eileen. 'Peasant life and rural conditions', c 1, 100–1, 500 (*Cambridge Medieval History* VII. 1932, p 729).

2 Singer, Charles. *A short history of science to the 19th century* (Oxford 1941, p 128).

3 Rück, Karl. 'Die Naturalis Historia des Plinius im Mittelalter' in *Sitz Berichte D. H. Bayer. Akad. d. Wiss* (Heft 2. intro 1898).
   Idem. 'Das Excerpt der Naturalis Historia des Plinius von Robert von Cricklade', *idem* (Heft 2. 1902).

4 Laistner, M. L. W. *Thought and letters in Western Europe AD 500–900* (2nd ed, 1957, p 57).

5 *Idem.* p 38.
   Taylor, H. O. *The classical heritage in the Middle Ages* (New York, p 112–13, 127–8).

6 James, M. R. 'Learning and literature to the death of Bede' (*Cambridge Medieval History* III. 1922, p 485–7).
   Roger, M. *L'enseignement des lettres classiques d'Ausone à Alcuin* (Paris 1905, p 171, 175).
   Taylor, H. O. op cit p 45.
   Idem. The medieval mind. 2 vols (1911 p 89).
   Putnam, G. Haven. *Books and their makers during the Middle Ages.* 2 vols (New York 1896, I. p 16–21).
   Lastri, Marco. *Biblioteca georgica ossia catalogo ragionato degli scrittori d'agricoltura* (Florence 1787, IX).
   Laistner. p 95–102.

7 Singer. p 128.
   Taylor. *Medieval mind* (p 103).
   Laistner. p 120.
   James. p 490.
   Lot, Ferdinand. *The end of the ancient world and the beginning of the Middle Ages* (1931 p 372).

8 Laistner. p 124

9 Roger. p 194–200.
   Taylor. *Classical heritage* (p 45).

10 Episcopi, Isidori Hispalensis. *Originum sive Etymologiarum* in *Auctores Latinae Linguae in unum redactie corpus* (1502 lib XIII).

11 Laistner. p 151, 154–8, 211.
    Roger. p 202, 289.
    Green, J. R. *A short history of the English people* (1907 p 38–41).
    Bolgar, R. R. *The classical tradition and its beneficiaries from the Carolingian age to the end of the Renaissance* (New York 1964, pp 105, 169).

12 Roger. p 440, 446.
    Laistner. p 229, 235.
    Putnam. p 62.

James. p 514.
Bolgar. p 107, 127–9.
Taylor. *Classical heritage* (p 219).
Easton, Stewart & Wieruszowski, Helene. *The era of Charlemagne. Frankish state and society* (New York 1961, p 95).

13 James. p. 522.
Rohde, Eleanor Sinclair. *The story of the garden* (1933 p 28–30).
Strabo, Walafrid. *Hortulus* tr Raef Payne. Commentary Wilfrid Blunt. (Hunt Botanical Library, Pittsburgh 1966, *passim*).

14 Amherst, Hon Alicia. *A history of gardening in England* (1895 p 4–5, c M. B. Guerard).

15 Laistner. p 191–4.
Fraas, C. *Geschichte der Landbau und Forstwirtschaft* (Munich 1865, p 24–6).
Dopsch, Alphons. *Economic and social foundations of European civilisation* (1953 p 329).
Guntz, Max. *Handbuch der landwirtschaftlichen Litteratur*. 2 vols (Leipzig 1897).

16 Bath, B. H. Slicher van. *The agrarian history of Western Europe* tr Olive Ordish (1963 p 40).
Latouche, Robert. *The birth of western economy* tr E. M. Wilkinson (1961 p 145).
Lizerand, Georges. *Le régime rural de l'ancienne France* (Paris 1942, p 5).

17 Fraas. p 25.

18 Stevens. C. E. 'Agriculture and rural life in the later Roman Empire' (*Cambridge Economic History*, 1941, I. p 91–5).

19 Laistner. p 170.
Taylor. *Classical heritage* (p 241, 247–8).
Bolgar. p 127.

20 Laistner, p 254–6.
Décarreaux, Jean. *Monks and civilisation* tr Charlotte Haldane (1964 p 240–50, 539–60).
Grimal, Pierre. *In search of ancient Italy* tr P. D. Cummins (1964 p 110–11).

21 Bath. *Agrarian history* (Table II. p 328–9).
Idem. 'Yield ratios 810–1820', *A. A. G. Bijdragen 10* (Wageningen 1963, p 30).

22 Usher, A. P. *A history of mechanical inventions* (1954, summing up many statements by archaeologists and others).

23 Examples are:
Boissonade, P. *Life and work in medieval Europe* tr Eileen Power (1927 p 25).
Crombie, A. C. *Augustine to Galileo. History of science AD 400–1650* (1952 p 16).

24 Wallace-Hadrill, J. M. *The long haired Kings, and other studies in Frankish history* (1962 p 3–9).
Lot, Ferdinand. *The end of the ancient world and the beginning of the Middle Ages* (Brussels 1931, p 365).
Moss, H. St L. B. *The birth of the Middle Ages* (Oxford 1935, p 245–6).
Church, R. W. *The beginning of the Middle Ages* (1895 p 63).

25 Grand, Roger. *L'agriculture au moyen âge de la fin de l'Empire Romain au XVIe siècle* vol III of *L'agriculture à travers les âges* ed Emile Savoy. (Paris 1940, p 50).

26 Wallace-Hadrill. p 2.
Prou, M. *La Gaule Merovingienne* (Paris nd, p 161–70).
Dill, Sir Samuel. *Roman society in Gaul in the Merovingian age* (1926 p 60, 68, 249).

27 Lizerand. p 28–31.
Seignobos, Charles. *The rise of European civilisation* tr C. A. Phillips (1939 p 53, 60).

28 Lot. p 308, 317, 368.
Grand. p 76.
Wallace-Hadrill. p 23.
Moss. p 244.
Boissonade. p 42–3.

29 Jullian, Camille. *Histoire de la Gaule.* 8 vols (Paris 1926, VIII p 164–5).

30 Schwerz, J. N. *Beschreibung der Landwirtschaft im Nieder-Elsass* (Berlin 1816, cap. XIII).

31 Stevens. op cit.
Boissonade. p 74.
Bloch, Marc. *Les caractères originaux de l'histoire rurale française* (Paris 1952, p 5).

32 Bath, B. H. Slicher van. *Agrarian history . . .* (p 66ff).

33 Ganshof. p 100–15.

34 Bloch. 'The rise of independent cultivation and seignorial institutions' (*Cambridge Economic History*, 1941, I. p 269–70).
Idem. Les caractères originaux (p 1–2, 5).
Moss. p 245.

35 Goltz, Theodor von der. *Geschichte der deutschen Landwirtschaft* (Stuttgart 1902–3, I. p 82–3).

Bury, J. B. *History of the later Roman Empire from Arcadius to Irene* (*AD 395–800*), 2 vols (1889, I. p 48).
Below, Georg von. *Geschichte der deutschen Landwirtschaft in ihre Grundzugen* (Jena 1937, p 44).

36 Seignobos. *Rise of European civilisation* (p 107–9).
Meitzen, August. 'Beobachtungen über Besiedelung, Hausbau und landwirtschaftliche Kultur', *Anleitung zur deutschends und volks Forschung* (Stuttgart 1889, p 496–8, 506, 510, 515).

37 Roupnel, Gaston. *Histoire de la campagne française* (Paris 1932, p 213).
East, Gordon. *An historical geography of Europe* (1935, 2nd ed 1943, p 57–8).
White, Lynn, Jr. *Medieval technology and social change* (Oxford 1962, p 69ff).

38 Goltz, von der. I. p 76.
Below, von. p 41–2 (Hoops. Lamprecht, Inama-Sternegg.)

39 Seignobos, Charles. *History of the French people* (1939 p 110–12).
Deansley, Margaret. *A history of early medieval Europe, AD 476–911* (p 142–3).
Crombie. p 16.
White, Lynn. p 41ff.
Moss. p 245–7 and many others.

40 Thirsk, Joan. 'The common fields', *Past and Present* (Dec 1964 p 8).
Baker, Alan R. H. 'Edward Levi Gray and English field systems' (*Agricultural History* 1965 XXXIX. p 89).

41 Collingwood, R. G. & Myres, J. N. L. *Roman Britain and the English settlements*. 2nd ed (Oxford 1937, p 27, 29)
Collingwood. *Roman Britain* (1937 p 77).
Clapham, J. H. *Concise economic history of Britain* (1949 p 18).
Vinogradoff, P. *Growth of the manor* (1905 p 182, 365).
Wood, P. D. 'Strip lynchets reconsidered' (*Geographical Journal* 1961, CXXVII. 4. p 453, 456–7).
Loyn, H. R. *Anglo-Saxon England and the Norman Conquest* (1962 p 147–7).

42 Blair, Peter Hunter. *An introduction to Anglo-Saxon England* (1956 p 270–1).
Idem. Roman Britain and early England, 55 BC– AD 871 (1963 p 259).
Stenton, F. M. *Anglo-Saxon England*. 2nd ed (1950 p 277, 282–3, 309).

43 Duby, Georges. *L'économie rurale et la vie des campagnes dans l'occident médiéval*. 2 vols (Paris 1962, I. p 74–9).
Abel, Wilhelm. *Geschichte der deutschen Landwirtschaft vom frühen Mittelalter bis zum 19 Jahrhundert* (Stuttgart 1962, p 18, 27).
Grand. p 272–5.

44 Bloch. *Cambridge Economic History* (1941. I).
Bath. *Agrarian history* . . . cap. III.
Abel. p 32.
Parain, Charles. 'The evolution of agricultural technique' (*Cambridge Economic History* 1941, I. p 127–8).

45 Heichelheim, Fritz M. 'Effects of classical antiquity on the land', *Man's role in changing the face of the earth* (Chicago 1956, p 178–9).

46 Lastri. IX.

47 Behlen, H. *Der Pflug und das Pflugen bei den Römern und in Mittel Europa in vorgeschichtlicher Zeit* (Dillenberg 1904, p 34).
Aslin, Mary S. *Library catalogue of printed books and pamphlets on agriculture published between 1471 and 1840* (Rothamsted, 2nd ed p 172).
Cripps-Day, Francis Henry. *The Manor Farm* (1931 p 21).

48 Guntz. I. p 53.

49 Russell, Bertrand. *History of western philosophy and its connection with political and social circumstances from the earliest times to the present day* (1947 p 444).
Glubb, John Bagot. *The Empire of the Arabs* (1963 p 327).
Taton, René. *A general history of science: ancient and medieval science from prehistory to AD 1450* tr A. J. Pomerans (1963 p 396).
Atiya, Aziz S. *Crusade, commerce and culture* (Indiana & Oxford UP 1962, p 209, 211–14).

50 Glubb. p 235.
East. p 189.
Way, Ruth & Simmons, Margaret. *A geography of Spain and Portugal* (1962 p 100–1).
Hitti, Philip K. *History of the Arabs*. 6th ed (1956 p 528).

51 Renan, Ernest. *An essay on the age and antiquity of the Book of Nabathean Agriculture* (1862 pref and p 7).
Huart, Clément. *A history of Arabic literature* (1903 p 316).
Hitti. p 352.
Cripps-Day. p 23.

52 Taton. p 400, 416.
Huart. p 316.

53 Clement-Mullet, J. J. tr of *Ibn al Awam* (p 68–79).
Cripps-Day. p 25–9.

54 Re, Filippo. *Elogio de Piero di Crescenzi detto nella Grand 'aula della Reale Universita da Bologna, Nov 1811* (Bologna 1812, p 7, 8).
Pouchet, F. A. *Histoire des sciences naturelles au moyen âge* (Paris nd).
Lacroix, Paul. *Science and literature in the Middle Ages and at the period of the Renaissance* (1878 p 114 and others).

55 Lastri. IX.

56 Pouchet. p 605.
    Jessen, Karl F. W. *Botanik der Gegenwart und Vorzeit in Culturhistorischer Entwicklung* reprint of 1864 ed. (Waltham, Massachusetts 1948, p 242).
    Re. *passim*.
    Goltz, von der. I. pp 290–1.
    Nisard, M. *Les agronomes latins* (1844 p 168).
    Goetz, Georgius (ed). *M. Terenti Varronis Rerum Rusticarum Libri Tres* (Teubner p v, xii).
    Madden. D. H. *A chapter of medieval history* (1924 p 95, 97).

57 Abel. p 149.

58 Lacroix. p 114.
    Storr-Best, Lloyd. (tr and ed) *Varro on farming* (1912 p xxxi).

59 Cited in Cripps-Day (p 4). See also Duby (I. p 311).

60 It would only multiply references if every detail were annotated. I do not therefore propose to do this.

61 Fussell, G. E. 'Soil classification in the 17th and 18th centuries', *Pochvovdeniya (Pedology)* no 5 (Moscow 1933).

62 Fussell, G. E. 'Marl, an ancient manure' (*Nature* 24 Jan 1959, p 214–17).
    Lizerand. p 5.
    Bloch. *Caractères originaux* (p 5).

63 Ed of 1548. Avicenna (980–1037) was an Arab writer on medicine, etc.

64 Abel. p 17, 36–8.

65 Reed, T. Dayrell. *The rise of Wessex* (1947 p 58–9).
    Singer. p 155.
    Taton. p 489–90.
    Taylor. *Medieval mind* (I. p 393 and others).

66 Steele, Robert. *Medieval lore from Bartholomew Anglicus* (1905 *passim*).

67 Pardi, Giulio del Pero. 'Gli attrezi da taglio per uso agricola in Italia dal antichita fino ad giorno nostri in relazione ad una falce con piccolo manico rappresentata in una miniatura di un codici manuscritto del XIV secolo esistente nel British Museum a Londra', *Nuovi annali dell'agricoltura* (Rome 1933, XIII. p 1–2).

68 Lindemans, Paul. *Geschiedenis van de landbouw in België*. 2 vols (Antwerp, II. p 59ff).
    Fussell, G. E. 'Hainault scythe in England' (*Man* no 144).

69 Bell, Claire Hayden. *Peasant life in old German epics* (Columbia Univ 1931, lines 814, 1057ff).

70 Eg, Lynn White. p 57ff.

71 Maitland, F. W. *Domesday Book and beyond* (1907 p 437–41).

72 Duby. I. p 311.

73 Bloch. *Caractères originaux* (p 5).
Parain. p 135.
Lizerand. p 5.

74 Steele. p 143.

75 Wright, Thomas (ed). *A volume of vocabularies from 10th to 15th centuries* (privately printed 1857).

76 Ernle, Lord. *English farming past and present*, with introductions by G. E. Fussell & O. R. McGregor. 6th ed (1961 p xlv–liv cap. I).

77 Eg, Slicher van Bath.

78 Pirenne, Henri. *A history of Europe* (1958 p 231–5).
Thorndike, Lynn. *Medieval Europe; its development and civilisation* (1920 p 290).
Boissonade. p 140 and others.

79 Power. p 725, 731.
Smith, L. M. *The early history of the monastery of Cluny* (Oxford 1920, pp 5, 6, 13).
Knowles, Dom David. *The monastic order in England . . . 940–1216* (Cambridge 1963, *passim*).
Décarreaux. p 358–9.
Church. p 62, 84, 144–5.

80 See Humphreys p 14–15 about bookshops and libraries in thirteenth-century Paris.

81 White, Lynn. cap II.
Taton. p 511.
Boissonade. p 157.

82 Power. p 731.

83 Hilton, R. H. 'Technical determinism: the stirrup and the plough', *Past and Present* (April 1963, no 24. p 95–100).
Abel. p 18, 39–40.
Duby. I. p 71ff.

84 Grand. p 272, 275–6. See plate 1 in Duby vol I.

85 Higgs, John. *The land* (1964 p 12).

Chapter Three

1 Schmitt, J. C. *Paladii Rutuli Tauri Aemiliani viri illustris opus Agriculturae* (Leipzig 1898, preface).

Notes

2 Green, J. R. *Short history of the English people* (ed 1907. p 298).
Thompson, James Westfall. *The medieval library* (Univ Chicago Studies in Library Science 1939, p 404–5).
Delisle, Leopold, *Recherches sur la libraire du Charles V* (Paris 1907, I. p 3, 4, 8, 33, 116).

3 Augé-Laribé, Michel. *La revolution agricole* (Paris 1955, p 14).

4 D'Israeli, Isaac. *Curiosities of literature.* 3 vols. 14th ed (1849, I. p 18–19).

5 Humphreys, K. W. *The library of the Carmelites of Florence at the end of the 14th century.* Studies in the history of libraries and librarianship. II. (Amsterdam 1964).

6 *Idem. The book provisions of the medieval friars, 1215–1400* (Amsterdam 1964, p 101–5, 112–14, 130).

7 Thompson. p 378, 381, 387; but see also:
Gibson, Strickland. *Some Oxford libraries* (Oxford UP 1914) whose index gives no reference to Cato, Columella, Varro, or astonishingly, Virgil.
Irwin, Raymond. *The English library* (1966 p 132).
Ker, N. P. *Medieval libraries of Great Britain. A list of surviving books* 2nd ed (Royal Historical Society 1964, p 22, 35, 42, 98, 168, 193, 208, 216).

8 Thompson. p 414, 425, 453.

9 Sandys, J. E. *History of classical scholarship from 6th century BC to the end of the Middle Ages* (Cambridge 1903, p 627–9).

10 Thompson. p 510–86.
Sabbadini, R. *Le scoperte dei Codici Latine e Greci ne'secolo XIV e XV* 2 vols. (Florence 1905. I. p 16, 25, 34, 74, 82, 87, 151, 184. II. p 14, 44, 50, 68, 74 and index).
Columella. Loeb ed (1955) ed Harrison Boyd Ash (I. p xvii).

11 Manitus, Max. *Handschriften antiker autoren in Mittelalterlichen Bibliotheks-katalogen* (Leipzig 1935, *passim*).
Haebler, Conrado. *Bibliografía Iberica del siglo XV. Enumeracion de todos los libros impressos en España y Portugal hasta el ano de 1500* (Leipzig 1903).

12 Kraus, Paul. *Jabir ibn Hayyan. Contribution à l'histoire des idées scientifiques dans l'Islam. vol II. Jabir et la science greque.* Mémoires de l'Institut d'Egypt 45 (Cairo 1942, p 79, 80, 84).
Bolgar. R. R. *The classical tradition and its beneficiaries* (New York 1964).

13 Augé-Laribé. p 14.

14 Schmitt. as note 1.
British Museum Catalogue.
Bibliographies in Loeb editions.

15 Sapori, Armando. *Studi di storia economica medievale* (Florence, 2nd ed 1946, p 707–8, 720–1).
    Hay, Denys. *The Italian renaissance in its historical background* (Cambridge UP 1966, p 54–66).
    Idem. Europe in the 14th and 15th centuries (1966 p 38).

16 This rough description follows the more detailed sketch made by P. J. Jones. 'The agrarian development of medieval Italy', *Second International Conference of Economic History* (Aix en Provence 1962, *passim*). It is confirmed by Gino Luzzato. *An economic history of Italy from the fall of the Roman Empire to the beginning of the 16th century* tr P. J. Jones (1961 p 157ff). See also:
    Herlihy, D. 'Population, plague and social change in rural Pistoia, 1201–1430', *Economic History Review* 2nd ser. XVIII. 2 (Aug 1965 p 231–43).
    Cipolla, C. M. 'The Italian and Iberian Peninsulas' (*Cambridge Economic History* III. 1963, p 409–11).
    Forti, Umberto. *Storia della technica dal medievo al rinascemento* (?1948 p 80–3, 90).

17 Sismondi, J. C. L. Sismondi di. *Historical view of the literature of the south of Europe.* 2nd ed (1846) tr Roscoe (I. p 246, 276–7, 304–11).
    Hay, Denys. *Italian renaissance* (p 119).
    Jerrold, Maud F. *Francesco Petrarcha, poet and humanist* (1909 p 49, 208, 210–11, 232–3).

18 Sismondi. p 313.
    Hay. *Ibid* p 121.

19 Wilkinson, L. P. 'The intention of Virgil's Georgics', *Greece and Rome.* XVIII. 55 (Jan 1950, p 19–21).

20 Power, Eileen. eg, 'Peasant life and rural conditions, c 1100–1500', *Cambridge Medieval History.* VII (1932 p 729–31).
    Bolgar. p 262–3.

21 Savage, Ernest A. *Old English libraries. The making, collection and use of books in the Middle Ages* (Antiquaries Books 1911, p 51–6, 63, 258. App B).

22 Even the general histories of England for the period confirm these conclusions, but they are far too numerous to be set out in a footnote; see, however, Maurice Beresford. *The lost villages of England* (1954 *passim*).

23 Boissonade, P. *Life and work in medieval Europe* tr Eileen Power (1927 p 316).
    Darby, H. C. 'The face of Europe on the eve of the great discoveries', *New Cambridge Modern History* I (1493–1520). (1957 p 25–7).
    Flores, Angel (ed). *Medieval age; specimens of European poetry from the 9th to the 15th centuries* (1965).

Perroy, Edouard. *Le moyen âge. L'expansion del'Orient et la naissance de la civilisation Occidentale* (Paris 1955, Tome III). *Histoire générale des civilisations* (p 403–8).

24 Abel, Wilhelm. *Geschichte der deutschen Landwirtschaft* (Stuttgart 1962, p 109–13, 115–18).
Fraas, G. *Geschichte der Landbau und Forstwirtschaft* (München 1865, p 31–2).

25 Seignobos, Charles. *The rise of European civilisation* tr C. A. Phillips (1939 p 208).
Bath, B. H. Slicher van. 'Economic and social conditions in the Frisian districts from 900 to 1500', *A. A. G. Bijdragen 13* (1965 p 125, 129–30).

26 Cannon, Grant C. *Great men of modern agriculture* (New York 1963, p 17).

Chapter Four

1 Fussell, G. E. 'A common market of ideas. The 16th century farming encyclopaedists', *Library Review.* 21. ii (Summer 1967, p 77–80).

2 Singer, Charles. *A short history of science to the nineteenth century* (Oxford 1941, esp chap VI).
Bush, Douglas. *Science and English poetry* (New York, Oxford UP 1950, p 5, 8, 17, 39).
Mousnier, Roland. *Les XVIe et XVIIe siécles. Progress of European civilisation and decline of the Orient, 1472–1715. Histoire générale des civilisations* vol IV (Paris 1954, p 2, 3, 10).
Jones, Richard Foster. *Ancients and moderns. A study of the rise of the scientific movement in 17th century England* (Washington Univ Studies, St Louis 1961, p 3, 4, 6, 21).
Digby, Kenneth. *Science, industry and social policy* (Edinburgh 1963, p 9, 11, 13).
Hall, A. Rupert. *From Galileo to Newton, 1630–1720. The rise of modern science* vol III (1963 p 175–6).
Simon, Joan. *Education and society in Tudor England* (Cambridge UP 1966, *passim*).
The volume of literature dealing with various aspects of the Renaissance and the Reformation is enormous in most languages: the above are some expressions of the modern outlook.

3 Aslin, Mary S. *Rothamsted Library Catalogue* 2nd ed (1940).
British Museum. *Library Catalogue.*
Guntz, Max. *Handbuch der landwirtschaftlichen Litteratur* 2 vol. (Leipzig, 1897 I. p 53, 75–6). Guntz regarded Crescentius, rightly, as the pattern and predecessor of the *Hausväterlitteratur.*
Fraas, C. *Geschichte der Landbau und Forstwirtschaft* (München 1865, p 31–3).
Hauser, Albert. 'Beitrage der Humanisten, insbesondere der Juristen zur

Entwicklung der Land und Forstwirtschaft vom 15 bis 17 Jahrhundert',
Zeitschrift für Agrargeschichte 14. 2 (Nov 1966, p 189).
Lastri, Marco. *Bibliotheca georgica ossia catalogo ragionato degli scrittori di
agricoltura* (Florence 1787, p 39, 74).

4 McDonald, Donald. *Agricultural writers, 1200–1800* (1908 p 9–12).
It is not necessary, but is perhaps excusable, to repeat that Cunningham
and Lamond edited Walter of Henley, Grosseteste, etc, in 1890, and
that the *Booke of Thrift* was reprinted in facsimile by Francis Henry
Cripps-Day in *The Manor Farm* (1931).
Bennett, H. S. *English books and readers, 1475–1557* (Cambridge UP 1952,
p 111–12).
Sawyer, Charles J. & Darton, F. J. Harvey. *English Books, 1475–1900*
(1927 I). Page 60 mentions Tusser but no other sixteenth-century agricultural writer.

5 Ernle, Lord. *English farming past and present*, with introductions by G. E.
Fussell & O. R. McGregor 6th ed (1961 p lvii–lxii, 90–1).
Thirsk, Joan. 'Farming techniques', *The agrarian history of England and
Wales*, vol IV. 1500–1640 (Cambridge UP 1967, p 169, 178, 187, 211).

6 Sells, A. Lytton. *The paradise of travellers. The Italian influence on Englishmen
in the 17th century* (1964 chapt I).
Carpenter, Rhys. *The Greeks in Spain* (1925 p 27–8).
Carrier, E. H. *Water and grass. A study in the pastoral economy of Southern
Europe* (1932 chapts X, XI).
Bath, B. H. Slicher van. *The agrarian history of western Europe, AD 500–
1850* tr by Olive Ordish (1963 p 167–8).
Salomon, Noël. *La campagne de la Nouvelle Castille à la fin du XVIe siècle
d'après les 'Relaciones topographicas'*, Les Hommes et la Terre, IX (Paris
1964, p 70–95).
Cf. Borrow, George. *The Bible in Spain* (1842).

7 Lastri. op cit. *passim*.
Moretti, K. *Bibliographia agronomica. Saggio d'un catalogo ragionato di libri
d'Agricoltura e Veterinaria scritti in Italiano* (Milan 1844).
Sismondi, J. C. L. Sismondi di. *Historical view of the literature of southern
Europe* tr by Roscoe. 2nd ed (1846 p 350).
Aslin. op cit.
Re, Filippo. *Saggio di bibliografia georgica* (Venice 1802). This is not in the
British Museum, and I have not been able to see it.

8 Montaigne. *The complete works of . . .* tr Wm Hazlitt, 'Diary of a journey
into Italy . . . 1580–1' (1842 p 560–613).
Bertagnolli, C. *Della vicenda dell'agricoltura in Italiae* (Florence 1881, p 224,
256, 258, 266, 315, 325).
Renard, G. & Weulersse, G. *Life and work in modern Europe* (1926 p 264,
267–9, 274, 276, 278).

Forti, Umberto. *Storia della tecnica dal medievo al rinascemento* (?1948 p 90, 92–4).

Luzzato, Gino. *An economic history of Italy from the fall of the Roman Empire to the beginning of the 16th century* tr Philip Jones (1961, p 101, 157, 166–7).

Lavan, Peter. *Renaissance Italy, 1464–1534* (1966 p 24–31).

Herlihy, David. *Medieval and renaissance Pistoia. The social history of an Italian town, 1200–1430* (Yale UP 1967, p 20, 37–44, 48).

9   Forti. *Idem* p 104–8.

Fussell, G. E. *The farmers' tools . . . AD 1500–1900* (1952 p 94–5).

Poni, Carlo. 'Richerche sugli inventori Bolognese della macchina semina-trice alla fine del secolo XVI', *Rivista Storica Italiana* LXXVI (1964 p 455–69, and the authorities cited therein).

10 Augé-Laribé, Michel. *La révolution agricole* (Paris 1955, p 14–15).

Lizerand, Georges. *Le régime rural de l'ancienne France* (Paris 1942, p 75).

11 Prothero, R. E. (Lord Ernle). *The pleasant land of France* (1908 p 33).

Lizerand. *Ibid* p 179–80.

12 Browne, Charles A. 'A source book of agricultural chemistry', *Chronica Botanica* VIII. i (Waltham, Massachusetts 1944, p 29–32).

Augé-Laribé. op cit p 15.

Rocque, Aurèle la. *The admirable discourses of Bernard Palissy* (Univ Illinois Press 1957, p v, vi, 5, 19, 127–35, 204–32).

13 Lizerand. op cit p 80.

Demolon, Albert. *L'évolution scientifique et l'agriculture française* (Paris 1946, p 10).

Villard, Eugène. *Olivier de Serres et son oeuvre* (Paris 1872).

Vaschelde, Henri. *Olivier de Serres, Seigneur de Pradel. Sa vie et ses travaux* (Paris 1886).

Pilon, Edmond. *Olivier de Serres* (Paris 1924).

Bibliothèque Nationale. *Les travaux et les jours dans l'ancienne France . . .* (1939).

Semple, Ellen Churchill. *Geography of the Mediterranean region; its relation to ancient history* (1932 p 302).

Grand, Roger. *L'agriculture au moyen âge de la fin de l'Empire Romain au XVIe siècle.* vol III. *L'agriculture à travers les âges* (Paris 1940, p 326–54 esp 336). Later in the book is an exhaustive discussion of livestock breeding, horses, donkeys, cattle, sheep and pigs.

NB. There are innumerable modern works on the farming, etc, of particular localities in the sixteenth century, and the systems described follow the usual pattern of difference between the open fields of the north and the enclosed areas, but this subject is best studied in one or more of the general works unless great detail is required.

14 Baudrillart, Henri. *Les populations agricoles de la France, 3rd ser. Du Midi*, etc (Paris 1893, p 56, 65–6, 250). He wrote other similar studies.

Notes

Seignobos, Charles. *History of the French people* (1939 p 226, etc).

15 Guntz, Max. op cit (I. p 108–9).
Goltz, Theodor von der. *Geschichte der deutschen Landwirtschaft.* 2 vols (Stuttgart 1901–2, I. p 291–5).
Fraas. op cit (p 37–41).
Renan, Ernest. *An essay on the age and antiquity of the Book of Nabathean agriculture* (1862 *passim*).
Schröder-Lembke, Gertrud. 'Die Hausväterlitteratur als Agrargeschichteliche Quelle', *Zeitschrift für Agrargeschichte* . . . (1953 p 109–19).
Abel, Wilhelm. *Geschichte der deutschen Landwirtschaft* (Stuttgart 1962, p 149ff).
Hauser, Albert. op cit (p 189–92).

16 Heresbach, Conrad. *Rei rusticae libri quatuor* . . . (1570 *passim*).
Goltz, von der. op cit I. p 294–5.
Schröder-Lembke, Gertrud. *Zwei frühe deutsche Landwirtschaftsschriften; Martin Grosser und Abraham von Thumbschirn* (Stuttgart 1965).
Idem. 'Die Genesis des Colerschen Hausbuches und die Frage seines Quellenwertes', *Wege und Forschungen der Agrargeschichte*. Essays presented to Günter Franz on his sixty-fifth birthday. (Frankfurt am Main 1965, *passim*).

17 Babel, Anthony. *Histoire économique de Genève des origines au début du XVIe siècle* (Geneva 1963, I. p 395–6. II. p 57–9).

18 Some thirty years or so ago, I published my conclusions on this subject as a signed back-page article in *The Times Lit Suppt*. Unfortunately, I have lost the reference and cannot find the cutting.

19 Thirsk, Joan. op cit. p 161–99.
Kerridge, Eric. *The agricultural revolution* (1967).
Fussell, G. E. 'Social and agrarian background of the Pilgrim Fathers' (*Agricultural History* 7. iv. Oct 1933, p 183–202).

20 Bennett, H. S. op cit. p 226–8.

21 Fuller, Thomas. *Worthies* ed of 1840. III (p 104).
Spencer, T. B. J. 'Lucretius and the scientific poem in English' in *Studies in Latin literature and influence* ed by D. R. Dudley (1965 p 153).

Chapter Five

1 The authorities upon which the bibliographical information is based are: The British Museum catalogue.
Aslin, Mary S. *The Rothamsted Experiment Station catalogue of books on agriculture, 1471–1840* (1940).
Fussell, G. E. *The old English farming books* . . . *1523–1730* (1947).
——. *Catalogue of the Walter Frank Perkins agricultural library* (Univ of Southampton 1961).

200

Notes

Lastri, Marco. *Biblioteca georgica ossia catalogo ragionato degli scrittori di agricoltura* (Florence 1787).

Re, Filippo. *Saggio di bibliographia georgica* (Venice 1802).

Moretti, K. *Bibliographia agronomica. Saggio d'un catalogo ragionato de Libri d'Agricoltura e Veterinaria scritto in Italiano* (Milan 1844).

Ramirez, Braulio Anton. *Diccionario de bibliografía agronomica* (Madrid 1865).

Hérissant, Louis Antoine Prosper. *Bibliothèque physique de la France ou liste de tous les ouvrages tant imprimé que manuscrits* . . . (on natural history of realm) (Paris 1771).

Musset-Pathay, V. D. de. *Bibliographie agronomique ou dictionnaire raisonné des ouvrages sur l'économie rurale et domestique* (Paris 1810).

Langethal, Chr Ed. *Geschichte der teutschen Landwirtschaft* (Jena 1854).

Fraas, C. *Geschichte der Landbau und Forstwirtschaft*, from sixth century to date (München 1865).

Guntz, Max. *Handbüch der landwirtschaftlichen Litteratur*. 2 vols (Leipzig 1897).

Poel, J. M. G. van der. *Wegwijzer in de Landbouwgeschiedenis* (Zwolle 1953).

2 Burckhardt, Jacob. *Judgments on history and historians* tr Harry Zohn (1959 p 31).

Highet, Gilbert. *The classical tradition. Greek and Roman influences on Western literature* (4th ed 1959. 1st 1949 p 262).

3 *The essays of Abraham Cowley* (ed 1869 p 39-47).

Hartlib, Samuel. *An essay for the advancement of husbandry learning . . . college of husbandry* (1651).

4 Clarke, M. L. *Classical education in Britain, 1500–1900* (Cambridge UP 1959, p 37).

Rhys, Hedley (ed). *Seventeenth century science and the arts*, a symposium (Princeton UP 1961, pref and p 19, 25).

Ogilvie, R. M. *Latin and Greek. A history of the influence of the classics on English life from 1600–1918* (1964 p 1, 37–8).

5 Quayle, Thomas. 'Gervase Markham, a reappraisal' (*Agriculture*, LXVI. 12).

Poynter, F. N. L. 'A bibliography of Gervase Markham, 1568(?)-1638' (*Oxford Bibliog Soc Publns NS*, XI. p viii, 218).

McDonald, Donald. *Agricultural writers, 1200–1800* (Field Office 1908, p 86).

6 Forbes, R. J. *Ancients and moderns. A study of the background of the battle of the books* (Washington Univ, St Louis 1936, p 18, 43, 125, 155, 159–63).

Idem. 'Food and drink', in vol III. Singer et al (*History of technology*, Oxford 1957, p 14–15).

Purver, Margery. *The Royal Society: concept and creation* (1967 p 20–42).

Notes

Note. The literature of this subject is very copious, and it would be irrelevant to cite more of it here.

7   Bandini, Mario. *Cento anni di storia agraria italiana* (Rome 1957, p 11).

8   Burnet DD, G. *Some letters containing an account of what seemed most remarkable in Switzerland, Italy, etc* (Rotterdam 1686, *passim*).

9   Sells, A. Lytton. *The paradise of travellers. The Italian influence on Englishmen in the 17th century* (1964 p 130, 160–9).
     Purver, Margery. op cit p 20.

10 Poni, Carlo. 'Ricerche sugli inventori Bolognesi della macchina seminatrice alla fine del secolo XVI', *Rivista Storica Italiana*, LXXVI. ii (1964 p 455–69).

11 Villard, Eugene. *Olivier de Serres et son oeuvre* (Paris 1872, p viii, 14, etc).
     Vaschelde, Henry. *Olivier de Serres. Sa vie et ses travaux* (Paris 1886).
     Pilon, Edmond. *Olivier de Serres* (Paris 1924, p 12, 24, 31).
     Bibliothèque Nationale. *Les travaux et les jours dans l'ancienne France* (Paris 1939, *passim* but esp p 12, 63ff).

12 Bloch, Marc. *Les caractères originaux de l'histoire rurale française* (Paris 1952, p 50, 218–21).
     Chavanne, C. Dareste de la. *Histoire des classes agricoles en France* 2nd ed (Paris. 1858, p 465, 470–2).
     Prothero, R. E. *The pleasant land of France* (1908 p 34–6).

13 Hérissant. op cit p 30.

14 Goltz, Theodor von der. *Geschichte der deutschen Landwirtschaft* 2 vols (Stuttgart. 1901–2. I. p 84).
     Small, Albion W. *The Cameralists. The pioneers of German social polity* (Univ Chicago Press 1909, p 5, 155, 187).

15 Abel, Wilhelm. *Geschichte der deutschen Landwirtschaft* (Stuttgart 1962, p 151–2, 185).
     Schröder-Lembke, Gertrud. 'Die Hausväterlitteratur als agrargeschichteliche Quelle', *Zeitschrift für Agrargeschichte* (p 111–13).
     Idem. Zwei frühe deutsche Landwirtschaftschriften; Anleitung zu der Landwirtschaft: Oeconomica (Stuttgart 1965).
     Idem. 'Die Genesis des Colerschen Hausbüches und die Frage seines Quellenwertes', *Wege und Forschungen der Agrargeschichte*. Essays presented to Günther Franz on his sixty-fifth birthday. (Frankfurt am Main 1967 *passim*).

16 Schröder-Lembke. 'Die Hausväterlitteratur' (as note 15, p 113–14).
     Abel. op cit p 188.
     Bruner, Otto. *Adeliges Landleben und Europäischer Geist. Leben und Werk Wolf Helmhardt von Hohberg, 1612–1688* (Salzburg 1949, p 294, 297–8).

Notes

Haushofer, Heinz. 'Der geistliche Einfluss der französischen auf die deutsche Landwirtschaft', *Zeitschrift für Agrargeschichte* (16. i. April 1968, p 2-3).

17 Hohberg, von. I. p iii and the bibliographers.

18 Renard & Weulersse. *Life and work in modern Europe* (1926 p 295).
Dickinson, Robert E. *Germany. A general and regional geography* (1953).
Mousnier, Roland. *Les XVIe et XVIIe siècles* (Progress of European civilisation and decline of Orient) *1472-1715. Histoire générale des civilisations.* IV (Paris 1954, p 145-7).
Murray, John J. 'The cultural impact of the Flemish Low Countries on 16th and 17th century England' (*American Historical Review.* LXII. 1957 p 851-3).
Fussell, G. E. 'Low Countries influence on English farming' (*English Historical Review.* Oct 1959 p 611-22).

19 Browne, Charles A. 'A source book of agricultural chemistry' *Chronica Botanica.* (Waltham, Massachusetts 8. i. Spring 1944 p 45ff).
Purver, Mary. *The Royal Society; concept and creation* (1967, *passim*).
Sells, A. Lytton. *The paradise of travellers* . . . (1964 p 12).
Ogilvie, R. M. op cit p 1, 4-6, 9, 22, 39.
Rhys Hedley (ed). op cit p 72, 81-2.
Nicholson, Marjorie. 'The microscope and English imagination' (*Smith College Studies in Modern Languages.* XVI. iv. 1934-5 p 5, 16, 22-3).
Casaubon, Meric (tr). *The golden book of Marcus Aurelius.* Intro W. H. D. Rouse (Everyman ed p 65).

Chapter Six

1 Bronson, Bertrand H. 'The writer', *Man versus Society in 18th century Britain* ed James L. Clifford (Cambridge UP 1968, p 106-7).

2 Ogg, David. *Europe of the Ancien Régime, 1715-1783* (Fontana History of Europe, 1967 p 308-10).

3 Nisard, M. *Les agronomes latins, avec le traduction en français* (Paris 1844).

4 Thomas, W. Hamshaw: in a lecture given to the British Agricultural History Society.

5 Bene, Benedetto del. *Opera di agricoltura* . . . *precedente dalla traduzione della Georgica di Virgilio* (Milan 1850, intro and p 95-6, 267).
Musset-Pathay, V. D. de. *Bibliothèque agronomique* . . . (Paris 1810).
Beckmann, Johann. *Grundsätze der teutschen Landwirtschaft* (Göttingen 1790, p 5).
Ramirez, Braulio Anton. *Diccionario de Bibliografía agronomica* (Madrid 1865, p 129).
Beckmann. p 3.

6 Tull, Jethro. *Horse hoeing husbandry.* 4th ed (1762 p 280–2. This the ed used by Dickson).

7 See Wilkinson, Patrick.

8 Abbe, Elfrida. *The plants of Virgil's Georgics* (Cornell UP 1965, p 117–24).

9 Benson, William. *Virgil's husbandry or an essay on the Georgics with Latin text and Dryden's translation* (1725).
Cf Edwards, Edward. *Libraries & founders of libraries* (1865, p 31), which mentions 70 collective and 25 partial eds.

10 Münchhausen, Otto von. *Der Hausväter . . .* 3 vols (Hanover 1765, p 374–6).

11 Switzer, Stephen. *Ichnographia Rustica, or the Nobleman, Gentleman and Gardener's Recreation.* 3 vols (1718, I. Pref, p iii, iv, xii, 30–2, 42).

12 Harte, Walter. *Essays on husbandry* (1764 *passim*).

13 Country Gentleman. *New system of agriculture.* 2nd ed (1755).
Maxwell of Arkland, Robert. *The practical husbandman* (Edinburgh 1757).
Hitt, Thomas. *A treatise on husbandry* (1760).
Mills, John. *A new and complete system of practical husbandry.* 2 vols (1762).
Forbes, Francis. *The modern improvements in agriculture.* new ed (1784).
Winter, George. *A new and comprehensive system of husbandry* (Bristol 1787).
Forsyth, Robert. *The principles and practice of agriculture systematically explained.* 2 vols (Edinburgh 1804).
This list is intended as an indication of the universality of the practice. It is not selected to support a case. The choice is quite haphazard.

14 Fussell, G. E. 'Technique of early field experiments' (*Journal of Royal Agricultural Society of England* 1935, 96, p 78–88).
Bourde, André. *Agronomie et agronomes en France au XVIIIe siècle.*, 3 vols (Paris 1967, I. p 213).

15 Treneer, Anne. *The mercurial chemist. A life of Sir Humphrey Davy* (1963 p 90).

16 Marshall, T. H. 'Jethro Tull and the new husbandry' (*Economic History Review* Jan 1929, p 41–60).

17 Brown of Markle, Robert. *Treatise on rural affairs.* 2 vols (Edinburgh. I. Intro and p 6–19).

18 A Practical Farmer (David Henry). *The complete English farmer* (1771 preface).

19 Bourde. op cit I. p 254.

20 *Ibid.* p 449.

Langethal, Chr Ed. *Geschichte der teutschen Landwirtschaft*. 4 vols (Jena 1854, IV. p 286–98, 317, 337).

21 Browne, Charles A. *A source book of agricultural chemistry. Chronica Botanica* (Waltham, Massachusetts 1944, I. p 126ff).
Musset-Pathay, de. op cit.
Müller, Hans Heinrich. 'Christopher Brown, an English farmer in Brandenberg, Prussia, in the 18th cent' (*Agricultural History Review*, 1969, 17 ii, p 120–135).
Schröder-Lembke, Gertrud. 'Englische Einflusse auf die deutsche Gutwirtschaft in 18 Jahrhundert', *Zeitschrift für Agrargeschichte* . . . (April 1964, 12. I. p 31).

22 Harte. op cit.

Bene, Benedetto del. *L'agricoltura di Lucio Giunio Moderato Columella* (Verona 1808, p viii).

23 Montelatici, Ubaldo. *Ragionamento sopra i mezzi piu necessari per far refiorire l'agricoltura* (Naples 1753, p 17 fn). This writer knew all the literature from classical and biblical times to Maggazine.
Griselini, Francesco. *Nuova maniera di seminare e coltivare il Formento*. 3rd ed (Venice 1765, *passim*). This book was also printed in Florence, 1764, and is attributed by the British Museum and Rothamsted to Giuseppe Rigacci, but he was the bookseller-publisher, not the author. The contents of both are identical.
Stancorich, Canon Pietor. *L'aratro-seminatore* (Venice 1820).
Campini, Antonio. *Saggi d'agricoltura* (Turin 1774), to which a specific discussion of the Norfolk system was added as well as a note about growing potatoes.

24 Forbes, R. J. *The conquest of nature. Technology and its consequences* (1968 p 27–8).
Cf Bronson, Bernard H. op cit. Commentary by James A. Clifford (p 151).

25 Armens, Sven M. *John Gay, social critic*. Reprint (Cass 1966, *passim*).
Thomson, James. *The Seasons* (1726).
Dyer, John. *The Fleece* (1757).
Parker, Edward A. & Williams, Ralph M. 'John Dyer, the poet, a farmer' (*Agric Hist USA* 22 July 1948, p 134–41).
Unwin, Rayner. *The Rural Muse. Studies in the peasant poetry of England* (Allen & Unwin 1954, p 25, 33, 35–7, 146).
It is noteworthy that Unwin makes the same quotations from Stephen Duck as Edmund Blunden in his *Nature in English Literature* (Hogarth Press 1929).
Hussey, Christopher. *The Picturesque* (1927).
Spencer, T. B. J. 'Lucretius and the scientific poem in English', *Studies in Latin literature and influence* ed by D. R. Dudley (1965 p 152).

26 Desplaces, L. B. *Préservatif contre l'agronomie ou l'agriculture réduit à ses vrais principes* (Paris 1762).

Patullo, M. *Essai sur l'amélioration des terres* (Paris 1758).

27 Bloch, Marc. *Les caractères originaux de l'histoire rurale française* (Paris 1952, p 222).
Duby, Georges & Mandrou, Robert. *A history of French civilisation from AD 1000 to the present day* tr James Blakeley Atkinson (1965, 1st 1958, p 391, 403).
Ducros, Louis. *French society in the 18th century* (1926).
Augé-Laribé, Michel. *La révolution agricole* (Paris 1955, p 53, 55–6–8, 65).
Dion, Roger. *Essai sur la formation du paysage rural françaisé* (Tours 1934, p 24, 77–9, 119).

28 Bourde. *Agronomes* (I. p 233–4).

29 Duby & Mandrou. op cit p 393ff.
Ducros, Louis. *Les Encyclopédists* (Paris 1900, p vi, 3, 9, 32, 47, 100 *et passim*).
Higgs, Henry. *The Physiocrats* (1897 Lectures I, II, III).
Renard, G. & Weulersse, G. *Le mouvement physiocratique en France.* (c 1908)

30 Bourde. *Influence* (p 22–5, 53, 56).
——. *Agronomes* (I. p 203, 242, 277, 302, 311).
Young, Arthur. *Travels in France*, ed Miss Betham Edwards (1905 p 137–9).

31 Abel, Wilhelm. *Geschichte der deutschen Landwirtschaft* (1962 p 185–7).
Krzymowski, R. *Geschichte der deutschen Landwirtschaft* (1939 p 159).
A list of some of these works follows; it does not pretend to be complete:
Becher, J. J. *Kluger Hausväter* (1702, 1709, 1714, 1764, 1800).
Atzendorf, Fischer von. *Hausväter, Gartner und Jägermeister* (1705).
Fischer, F. A. *Hausväter oder fleissiges Herrenauge* (1696, 1750).
Florini, Francisci Phillippi (Pfalzgraf Franz Phillip bei Rhein). *Allgemeiner klug und rechtverstandiger Hausväter* (1702).
Idem. Oeconomus prudens et legalis . . . oder grosser Herren Stands und Adelicher Hausvater (at least 4 eds to 1751).
König. *Georgica helvetica curiosa* (1706).
Rohr, Julius Bernhard von. *Compendieuse Haushaltungs Bibliothek* (1716), and other works.
Feldeck, Joseph von. *Haus und Landwirtschaft* (1718).
Liberti. *Churfächsische adelige Hauswirtschaftkunst* (1701).
Furstenau. *Anleitung zu Haushaltung und Viehartznei-Kunst* (1736).
Hoffman, Gottfried August. *Klugheit Hauszuhalten.* 5 vols (several eds to 1742).
Schröder. *Landbauer* (1713).
Hellwig, L. Christ von. *Haus, Feld und Arztniebuch* (1718 etc).

32 Abel. op cit p 258–60.
Small, Albion W. *The Cameralists. The pioneers of German social polity* (Chicago UP 1909, p 207, 230–3).

33 Langethal, Chr Ed. *Geschichte der teutschen Landwirtschaft* 4 vols (Jena, 1854, IV. p 286, 288).
Fraas, C. *Geschichte der Landbau und Forstwirtschaft seit den 16 Jahrhundert bis zur Gegenwart* (München 1865, p 86).

34 Gunst, Max. *Handbüch der Landwirtschaftlichen Literattur* (1897).
Boelke, Willi Alfred. 'Max Gunst, Landwirt und Publizist', *Zeitschrift für Agrargeschichte* . . . (12. I. April 1964, p 54–65 is an appreciation of Gunst).
Goltz, Theodor von der. *Geschichte der deutschen Landwirtschaft.* 2 vols (ed 1963, I. p 358ff, 450).

35 Augé-Laribé. op cit *passim.*
Gromas, Raymond. *Histoire agricole de la France des origines à 1939* (Mende 1947, p 62ff).
Chambers, J. D. & Mingay, G. E. *The agricultural revolution, 1750–1880* (1966).
Kerridge, Eric. *The agricultural revolution* (1967).
Jones, E. L. (ed). *Agriculture and economic growth in England, 1650–1815* (Univ paperback 231. 1967).

36 Haushofer, Heinz. 'Der geschichteliche Einfluss der französischen auf die deutsche Landwirtschaft', *Zeitschrift für Agrargeschichte* (16. i. April 1968, p 5–7).

37 Schröder-Lembke, Gertrud. 'Englische Einfluss auf die deutsche Gutswirtschaft im 18 Jahrhundert.' *Zeitschrift für Agrargeschichte.* (12. I. April 1964 p 29–36).
Thaer is discussed in all the German histories of agriculture as might be expected. Refs are unnecessary here.

38 See my 'Eighteenth century theories of crop nutrition' (not yet printed). A paper read at the annual conference of the British Agricultural History Soc at East Anglian Univ. April 1968.

39 Prato, Giuseppe. *L'evoluzione agricola nel secolo XVIII* (1909 p 73).
——. *La vita economica in Piemonte a mezzo il secolo XVIII* (Turin 1908, p 66, 69, 70, 81).
Pugliese, Salvatore. *Due secolo di vita agricola* (1908 *passim*). Pugliese estimates yields rather higher than Prato.
Woolf, S. J. 'Economic problems of the nobility in the early modern period: the example of Piedmont' (*Econ Hist Rev*, 2nd ser. XVII. 2. Dec 1964 p 263–70).
Addison, Joseph. *Remarks on several parts of Italy etc in the years 1701, 1702, 1703* (1705), gives an intelligent traveller's observations on farming.

40 Poni, Carlo. *Gli aratri e l'economia nel Bolognese dall' XVII ad XIX secolo* (Bologna 1963, p 4, 10, 14, 15, 17).

41 Bandini, Mario. *Centi anni di storia agraria italiana* (Rome 1957, p 11, 12).

42 Ramirez, Braulio Anton. *Diccionario de bibliographia agronomica* (Madrid 1865).

43 Poel, J. M. G. van der. *Heeren en Boeren. Een studie over de Commission van Landbouw, 1805–1851* (Wageningen 1949, p 22 fn).
Idem. 'Beoefening van de landbouwgeschiedenis in het binnen- en buiten-land'. (*Landbouwgeschiedenis* 1960).

Chapter Seven

1 Huxley, Aldous. *Do what you will. Essays* (Thinker's Lib no 56, 1937 p 12, 23). Innumerable historiographers have dealt with this problem.

2 Thackeray, W. M. *The English humourists of the 18th century* (ed. nd. p 54). Steggall, John H. *A real history of a Suffolk man* ed by Richard Cobbold (c 1850 p 98).

3 Gregoire, G. *Essai historique sur l'état de l'agriculture en Europe au 17e siècle* (Paris An XII. p 4).

4 Bourde. *Agronomie* (I. p 447–8).

5 Highet, Gilbert. *The classical tradition. Greek and Roman influences on western literature* (Oxford paperbacks, 1967, 1st 1949. p 262, 288).
Glacken, Clarence J. *Traces on the Rhodian shore. Nature and culture in western thought from ancient times to the end of the 18th century* (Univ Calif Press 1967, p 131, 137).

6 Schultz, Theodore W. *Economic crises in world agriculture* (Univ Michigan Press 1965, p 21–2).

7 Sturgess, R. W. 'The agricultural revolution on the English clays' (*Agric Hist Rev.* XIV. 2. 1966 p 106–7).

8 White, R. J. *Europe in the 18th century* (1965 p 42–3).
Mumford, Lewis. *Tecnics and civilisation* (1947 p 146–7).
Ogg, David. *Europe of the Ancien Régime, 1715–1783* (1967 p 349).

9 Singer, Charles. *A short history of science to the 19th century* (1941 p 154–8).
Pollard, Sidney. *The idea of progress. History and society* (New Thinkers Lib, Watts 1968, p 7–9, 14).
Debus, Allen G. *The chemical dream of the Renaissance*, Churchill Coll Overseas Fellowship Lecture no 3 (Heffer 1968, p 8–10, 33).
Fussell, G. E. 'The technique of early field experiments' (*Jour Roy Agric Soc Eng.* 96. 1935 p 1–11).

10 Brower, Reuben Arthur. *Alexander Pope. The poetry of allusion* (1959 p 41, 49, 107, 113).
Gay, John. *The shepherds week in six pastorals* (1714, Oxford 1926).
Deane, C. V. *Aspects of 18th century nature poetry* (1st 1935, reptd 1967).

McKillop, Alan Dugald. *The background of Thomson's 'Seasons'* (Univ Minnesota Press 1942, p 43, 45–7, 50, 55).

Sambrook, A. J. 'The farmer's boy: Robert Bloomfield, 1766–1823' (*English* XVI. Summer 1967, no 95, p 167–70).

Clifford, James L. *Man and society in 18th century Britain* (Cambridge UP 1968), and, of course, the writings of the poets themselves. Innumerable critical works and appreciations have been written.

11 Galbraith, John Kenneth. *The new industrial state* (1967 p 52, cf p 201, 209).

Pollard, Sidney. op cit p 64.

12 Mumford, Lewis. *The human prospect* (1936 p 16).

Duby & Mandrou. op cit p 352–4, 387 fn.

Bate, Walter Jackson. *From classic to romantic. Premises of taste in 18th century England* (Harper Torchbooks 1036 1961, p 93).

Jones, William Powell. *The rhetoric of science. A study of scientific ideas and imagery in 18th century English poetry* (1966 p 5, 17, 79, 160, 201–2).

13 Forbes, R. J. *The conquest of nature. Technology and its consequences* (1968 p 21, 25, 43).

14 Daumas, Maurice. 'Precision of measurement and physical and chemical research in the 18th century', *Scientific Change* ed A. C. Crombie (1963, *passim*).

Fussell, G. E. *The farmers' tools . . . AD 1500–1900* (1952 p 40ff).

15 Brown, James Campbell. *A history of chemistry from the earliest times to the present day* (1913 p 450–6).

There are several other works that could be quoted. Indeed I hope that I may return to this subject in a full-scale history.

Bibliography

I

Abbe, Elfrida. *The plants of Virgil's Georgics* (1965).

Abel, Wilhelm. *Geschichte der deutschen Landwirtschaft von frühen Mittelalter bis zum 19 Jahrhundert* (1962).

Addison, Joseph. *Remarks on several parts of Italy . . . in the years 1701, 1702, 1703* (1705).

Amherst, Hon Alicia. *A history of gardening in England* (1895).

Anderson, I. *A history of Sweden* tr G. Caroline Hannay (1956).

Anon. 'On the agriculture of the Romans' (*Quarterly Journal of Agriculture* II. 1831).

Applebaum, Shimon. 'Agriculture in Roman Britain' (*Agricultural History Review* VI 1958).

Armens, Sven M. *John Gay, social critic* (Reprint, 1966).

Aslin, Mary S. *Rothamsted Library. Catalogue of printed books and pamphlets on agriculture published between 1471 and 1940* (2nd ed 1940).

Atiya, A. S. *Crusade, commerce and culture* (1962).

Augé-Laribé, Michel. *La révolution agricole* (1955).

Ausonius. *Poems* tr H. G. Evelyn White (Loeb Classical Library 2 vols).

Babel, Anthony. *Histoire économique de Genève des origines au début du XVIe siècle* (1963).

Baker, A. R. H. 'Howard Levi Gray and English field systems' (*Agricultural History* USA 39, 1965).

Bandini, Mario. *Cento anni di storia agraria italiana* (1957).

Bate, Walter Jackson. *From classic to romantic. Premises of taste in 18th century England* (1961).

Bateson, Mary. *Medieval England, 1066–1350* (1903).

Bath, B. H. Slicher van. *Agrarian history of western Europe* tr Olive Ordish (1963).

Idem. 'Yield ratios, 810–1820' (*A. A. G. Bijgragen 10* 1963).

Bibliography

Idem. 'Economic and social conditions in the Frisian districts from 900 to 1500' (*idem 13* 1965).

Baudrillart, Henri. *Les populations agricoles de la France. 3rd ser du Midi* (1893). This author wrote on many other parts of France.

Baumeister, A. *Denkmaler des klassichen Altertümer* (1884).

Beckmann, Johann. *Grundsätze der teutschen Landwirtschaft* (1790).

Behlen, H. *Der Pflug und das Pflugen bei den Römern und in Mittel Europa in vorgeschichtlicher Zeit* (1904).

Bell, C. H. *Peasant life in old German epics* (1931).

Below, Georg von. *Geschichte der deutschen Landwirtschaft des Mittelalters in ihren Grundzugen* (1937).

Bene, Benedetto del. *L'agricoltura di Lucio Giunio Moderato Columella* (1808).

Idem. Opera di agricoltura . . . precedenta dalla traduzione della Georgica di Virgilio (1850).

Bennett, H. S. *English books and readers, 1475–1557* (1952).

(Benson, William.) *Virgil's husbandry or an essay on the Georgics with Latin text and Dryden's translation* (1725).

Beresford, Maurice. *The lost villages of England* (1964).

Bertagnolli, C. *Della vicenda dell'agricoltura in Italia* (1881).

Bibliothèque Nationale. *Les travaux et les jours dans l'ancienne France* (1939).

Blair, P. H. *An introduction to Anglo-Saxon England* (1956).

Idem. Roman Britain and early England, 55 BC–AD 871 (1963).

Bloch, Marc. 'The rise of independant cultivation and seignorial institutions' (*Cambridge Economic History* I 1941).

Idem. Les caractères originaux de l'histoire française (1952).

Idem. Feudal Society tr L. A. Marryon (1961).

Blunden, Edmund. *Nature in English literature* (1929).

Boelke, Willi Alfred. 'Max Gunst, Landwirt und Publizist' (*Zeitschrift für Agrargeschichte* XII April 1964).

Boissier, Gaston. *The country of Horace and Virgil* tr D. Havelock Fisher (1896)

Boissonade, P. *Life and work in medieval Europe* tr Eileen Power (1927).

Bolgar, R. R. *The classical tradition and its beneficiaries from the Carolingian age to the end of the Renaissance* (1964).

Borrow, George. *The Bible in Spain* (1842).

Bourde, André. *The influence of England on the French agronomes, 1750–1789* (1953).

Bibliography

Idem. Agronomie et agronomes en France au XVIIIe siècle 3 vols (1967).

Brehaut, Ernest (trs). *Cato the Censor on farming* (1933).

Brentano, Lujo. *Das Wirtschaftsleben der antiken Welt* (1961).

Bronson, Bertrand H. 'The writer', *Man versus Society in 18th century Britain* ed James L. Clifford (1968).

Brower, Reuben Arthur. *Alexander Pope. The poetry of allusion* (1959).

Brown, James Campbell. *A history of chemistry from the earliest times to the present day* (1913).

Brown of Markle, Robert. *Treatise on rural affairs* 2 vols (1811).

Browne, Charles A. 'A source book of agricultural chemistry', *Chronica Botanica* VIII (1944).

Burn, Andrew Robert. *The world of Hesiod* (1936).

Bury, J. B. *History of the later Roman Empire from Arcadius to Irene, AD 395–800* 2 vols (1889).

Bush, Douglas. *Science and English poetry* (1950).

Campini, Antonio. *Saggi d'agricoltura* (1774).

Cannon, Grant. *Great men of modern agriculture* (1963).

Carpenter, Rhys. *The Greeks in Spain* (1925).

Carrier, E. H. *Water and grass. A study in the pastoral economy of southern Europe* (1932).

Cary, M. 'The general condition of Greece in 386 BC' (*Cambridge Ancient History* VI. 1927).

Cato. *de agri cultura* tr W. D. Hooper & H. B. Ash (Loeb Classical Library 1959).

Cato & Varro. *De re rustica* tr *idem idem* (1960).

Chambers, J. D. & Mingay, G. E. *The agricultural revolution, 1750–1880* (1966).

Chilver, G. E. F. *Cisalpine Gaul: social and economic history from 49 BC to the death of Trajan* (1941).

Church. R. W. *The beginnings of the Middle Ages* (1895).

Cipolla, C. M. 'The Italian and Iberian Peninsulas' (*Cambridge Economic History* III. 1963).

Clark, Colin & Haswell, M. *The economics of subsistence agriculture* (1964).

Clapham, J. H. *Concise economic history of Britain* (1949).

Clarke, R. Raynbird. *East Anglia* (1960).

Bibliography

Collingwood, R. G. & Myres, J. N. L. *Roman Britain and the English settlements* (2nd ed 1937).

Collingwood, R. G. *Roman Britain* (1937).

Columella. *Of husbandry in 12 books tr into English* (1745).

Idem. ed Harrison Boyd Ash (Loeb Classical Library, 3 vols 1955).

Coulton, G. G. *Medieval Village* (1925).

Country Gentleman. *New system of agriculture* (2nd ed 1755).

Crescentius, Petrus. *Librum commodorum ruralium* (? 1480).

Idem. Le bon mesnaiger (1540).

Idem. De omnibus agriculturae partibus (1548).

Cripps-Day, F. H. *The manor farm* (1931).

Crombie, A. C. *Augustine to Galileo. History of science AD 400–1650* (1952).

Idem. Robert Grosseteste and the origins of experimental science (1953).

Cunningham, W. & Lamond, Elizabeth (eds). *Walter of Henley's Husbandry together with an anonymous husbandry, Seneschaucie and Robert Grosseteste's Rules* (1890).

Darby, H. C. 'The face of Europe on the eve of the great discoveries' (*New Cambridge Modern History. I. 1493–1520* 1957).

Daumas, Maurice. 'Precision of measurement and physical and chemical research in the 18th century', *Scientific Change* ed A. C. Crombie (1963).

Deane, C. V. *Aspects of 18th century nature poetry* (1935).

Deansley, Margaret. *A history of early medieval Europe, AD 476–911* (1956).

Debus, Allen G. *The chemical dream of the Renaissance* (1968).

Décarreaux, J. *Monks and civilisation* tr Charlotte Haldane (1964).

Delisle, Leopold. *Recherches sur la librarie de Charles V* (1907).

Demolon, Albert. *L'évolution scientifique et l'agriculture française* (1946).

Derry, T. K. & Williams, Trevor I. *A short history of technology to AD 1900* (1960).

Desplaces, L. B. *Preservatif contre l'agronomie ou l'agriculture réduit à ses vrais principes* (1762).

Digby, Kenneth. *Science, industry and social policy* (1963).

Dill, Sir Samuel. *Roman society in Gaul in the Merovingian age* (1926).

Idem. Roman society from Nero to Marcus Aurelius (1957).

Bibliography

Dion, Roger. *Essai sur la formation du paysage rural français* (1934).

D'Israeli, Isaac. *Curiosities of literature* (14th ed 1849).

Dopsch, Alphons. *Economic and social foundations of European civilisation* (1953).

Duby, Georges. *La société aux XIe et XIIe siècle dans la région Mâconnaise* (1953).

Idem. L'économie rurale et la vie des campagnes dans l'occident mediéval 2 vols (1962).

Idem et Mandrou, Robert. *A history of French civilisation from AD 1000 to the present day* tr James Blakeley Atkinson (1965, 1st 1958).

Duckett, E. S. *Latin writers of the fifth century* (1930).

Idem. Carolingian portraits (1962).

Ducros, Louis. *Les Encyclopédists* (1900).

Idem. French society in the 18th century (1926).

Dyer, John. *The fleece* (1757).

East, G. *An historical geography of Europe* (2nd ed 1943, 1st 1935).

Easton, S. & Wieruszowski, Helene. *The era of Charlemagne. Frankish state and society* (1961).

Edwards, Edward. *Libraries and founders of libraries* (1865).

Ernle, Lord. *English farming past and present*. With introductions by G. E. Fussell & O. R. McGregor (6th ed 1961).

Evans, Joan. *Life in medieval France* (1959, 1st ed 1925).

Eyre, S. R. *Vegetation and soils. A world picture* (1963).

——. *Fleta* vol II. 1955, ed with a translation by H. G. Richardson & G. O. Sayles (*Selden Society* vol LXXII. 1953).

Flores, Angel (ed). *Medieval age: Specimens of European poetry from the ninth to the fifteenth centuries* (1965).

Foord, E. *The last age of Roman Britain* (1925).

(Forbes, Francis). *The modern improvements in agriculture* (new ed 1784).

Forbes, R. J. *The conquest of nature. Technology and its consequences* (1968).

Forsyth, Robert. *The principles and practice of agriculture systematically explained* 2 vols (1804).

Forti, Umberto. *Storia della tecnica dal medievo al rinascemento* (? 1948).

Fowler, J. *Medieval Sherborne* (1951).

Fraas, C. *Geschichte der Landbau und Forstwirtschaft* (1865).

Bibliography

Fuller, Thomas. *Worthies* 3 vols (ed of 1840).

Fussell, G. E. 'Population and wheat production in the 18th century' (*History Teachers Miscellany* May 1929).

Idem. 'Soil classification in the 17th and 18th centuries', *Pochvovdeniya (Pedology)* no 5 (1933).

Idem. 'Social and agrarian background of the Pilgrim Fathers' (*Agricultural History* VII. 4. 1933).

Idem. 'Technique of early field experiments' (*Journal Royal Agricultural Society England* 96. 1935).

Idem. The farmers' tools . . . AD 1500-1900 (1952).

Idem. 'Marl: an ancient manure' (*Nature* 24 Jan 1959).

Idem. 'Hainault scythe in England' (*Man* no 144, 1960).

Idem. 'A common market of ideas. The 16th century farming encyclopaedists' (*Library Review* 21. II. Summer 1967).

Idem. 'Eighteenth century theories of crop nutrition'. Paper read at the annual conference of the British Agricultural History Society, April 1968—*not yet printed.*

Galbraith, John Kenneth. *The new industrial state* (1967).

Ganshof, F. L. *La Belgique Carolingienne* (1958).

Gasquet, A. *English monastic life* (1904).

Gay, John. *The shepherds week in six pastorals* (1714).

——. *Geoponika. Agricultural pursuits* tr from Greek by Rev T. Owen (1805).

Gibson, Strickland. *Some Oxford libraries* (1914).

Glacken, Clarence J. *Traces on the Rhodian shore. Nature and culture in western thought from ancient times to the end of the 18th century* (1967).

Glotz, Gustave. *Ancient Greece at work* (1926).

Glover, T. R. *Studies in Virgil* (1904). Citing Saint Beuve *Étude sur Virgile.*

Glubb, J. B. *The Empire of the Arabs* (1963).

Goetz, G. (ed). *M. Terenti Varronis Rerum Rusticarum Libri Tres* (1929).

Goltz, Theodor von der. *Geschichte der deutschen Landwirtschaft* 2 vols (1902-3).

Grand, R. *L'agriculture au moyen âge de la fin de l'Empire Romain au XVIe siècle* vol II of *L'agriculture à travers les âges* ed Emile Savoy (1940).

Green, J. R. *A short history of the English people* (ed of 1907).

Gregoire, G. *Essai historique sur l'état de l'agriculture en Europe au 17e siècle* An XII.

Bibliography

Grimal, P. *In search of ancient Italy* tr P. D. Cummins (1964).

Griselini, Francesco. *Nuove maniera di seminare e coltivare il Formento* (3rd ed 1765).

Gromas, Raymond. *Histoire agricole de la France des origines à 1939* (1947).

Guérard, M. B. (ed). *Polyptyque de St Irminion* (1844).

Guntz, Max. *Handbuch der landwirtschaftlichen Litteratur* 2 vols (1897).

Hadzits, George Dupue. *Lucretius and his influence* (1935).

Haebler, Conrad. *Bibliografía Ibérica del siglo XV. Enumeración de todos los libros impresos en España y Portugal hasta el ano de 1500* (1903).

Hall, A. Rupert. *From Galileo to Newton, 1630–1720. The rise of modern science* vol III (1963).

Harden, D. B. (ed). *Dark Age Britain. Studies presented to E. T. Leeds* (19 ?)

Harte, Walter. *Essays on husbandry* (1764).

Hartmann, L. M. *Zur wirtschaftliche Geschichte Italiens im frühen Mittelalter* (1904).

Hauger, Alphons. *Zur romischen Landwirtschaft und Haustierzucht* (1921).

Hauser, Albert. 'Beitrage der Humanisten, inbesondere der Juristen zur Entwicklung der Land und Forstwirtschaft' (*Zeitschrift für Agrargeschichte* 14. 2. Nov 1966).

Haushofer, Heinz. 'Der geschichliche Einfluss der französischen auf die deutschen Landwirtschaft' (ibid XVI. April 1968).

Hausväterlitteratur. 18th century. see note 31, chap 6 select list.

Hawkes, Christopher. 'The Roman villa and the heavy plough' (*Antiquity* IX. 1935).

Hay, Denys. *The medieval centuries* (2nd ed 1964).

Idem. Italian Renaissance in its historical background (1966).

Idem. Europe in the 14th and 15th centuries (1966).

Hayward, Richard Mansfield. *The myth of Rome's fall* (1958).

Heer, Fr. *The medieval world. Europe from 1100 to 1350* tr Janet Sondheomer (1962).

Heichelheim, F. M. 'Effects of classical antiquity on the land', *Man's role in changing the face of the earth* (1956).

Heitland, W. E. *Agricola: a study of agriculture and rustic life in the Graeco-Roman world* (1921).

Heresbach, Conrad. *Rei rusticae libri quattuor* (1570).

Bibliography

Herlihy, D. 'Population, plague and social change in rural Pistoia 1201–1430 (*Economic History Review* 2nd ser. XVIII. 2. Aug 1965).

Idem. Medieval and renaissance Pistoia. The social history of an Italian town, 1200–1430 (1967).

Hesiod. *Works and Days* tr Hugh Evelyn White (Loeb Classical Library 1954).

Higgs, Henry. *The physiocrats* (1897).

Higgs, J. *The land* (1964).

Highet, Gilbert. *The classical tradition. Greek and Roman influences on western literature* (1967, 1st 1949).

Hilton, R. H. 'Technical determinism: the stirrup and the plough' (*Past and Present* no 24, 1963).

Hitt, Thomas. *A treatise on husbandry* (1760).

Hitti, Philip K. *History of the Arabs* (6th ed 1956).

Homans, G. C. *English villages in the 13th century* (1942).

Honigsheim, Paul. 'Max Weber as historian of agriculture and rural life' (*Agricultural History* XXIII. July 1949).

Huart, C. *A history of Arabic literature* (1903).

Hubert, Henri. *The greatness and decline of the Celts* (1934).

Humphreys, K. W. *The book provisions of the medieval friars, 1215–1400* (1964).

Idem. 'The library of the Carmelites of Florence at the end of 14th century' (*Studies in the history of libraries and librarianship* II. 1964).

Hunger, F. W. T. *The herbal of Pseudo-Apuleius from a 9th century MS in the Abbey of Monte Cassino* (1935).

Hussey, Christopher. *The picturesque* (1927).

Huxley, Aldous. *Do what you will. Essays* (1937).

Ibn al Awam. *Le livre de l'agriculture (Kitab al Felahah)* tr J. J. Clement-Mullet (1864).

Irwin, Raymond. *The English library* (1966).

Isidori, Hispalensis Episcopi. *Originum sive Etymologiarum* in *Auctores Latinae Linguae in unum redactie corpus* (1502).

Jacks, L. V. *Xenophon, soldier of fortune* (1930).

James, M. R. 'Learning and literature to the death of Bede' (*Cambridge Medieval History* III. 1922).

Jerrold, Maud F. *Francesco Petracha, poet and humanist* (1909).

Bibliography

Jessen, K. F. W. *Botanik der gegenwart und Vorzeit in Culturhistorischer Entwicklung* (1948, reprint of 1864 ed).

Jones, E. L. *Agriculture and economic growth in England, 1650-1815* (1967).

Jones, P. J. 'The agrarian development of medieval Italy' (*Second International Conference of Economic History* 1962).

Jones, Richard Foster. *Ancients and moderns. A study of the rise of the scientific movement in 17th century England* (1961).

Jones, William Powell. *The rhetoric of science. A study of scientific ideas and imagery in 18th century English poetry* (1966).

Jope, E. M. 'Agricultural implements' (*History of Technology* II. 1956).

Jullian, C. *Histoire de la Gaule* 8 vols (1926).

Ker, N. R. *English manuscripts in the century after the Norman Conquest* (1960).

Idem. Medieval libraries of Great Britain; a list of surviving books (2nd ed 1964).

Kerridge, Eric. *The agricultural revolution* (1967).

Knowles, Dom D. *The monastic order in England, 940-1216* (1963).

Kraus, Paul. *Jabir ibn Hayyan: Contribution à la histoire des idées scientifiques dans l'Islam* (Mémoires de l'Institut d'Egypt no 45, 1942).

Krzymowski, R. *Geschichte der deutschen Landwirtschaft* (1939).

Lacroix, P. *Science and literature in the Middle Ages and at the period of the Renaissance* (1878).

Laistner, M. L. W. *Thought and letters in western Europe, AD 500-900* (1957).

Langethal, Chr Ed. *Geschichte der teutschen Landwirtschaft* 4 vols (1854).

Lastri, M. *Biblioteca georgica ossia catalogo ragionato degli scrittori d'agricoltura* (1787).

Latouche, Robert. *The birth of western economy: economic aspects of the Dark Ages* tr E. M. Wilkinson (1961).

Lattin, Harriet Pratt (tr). *The letters of Gerbert* (1961).

Lavan, Peter. *Renaissance Italy, 1464-1534* (1966).

Lefèbre, des Noëttes. *L'attelage. Le cheval de selle à travers les âges* (1931).

Lindemans, P. *Geschiedenis van de landbouw in België* 2 vols (1952).

Lizerand, Georges. *Le régime rural de l'ancienne France* (1942).

Lot, Ferdinand. 'Le jugum, le manse et les exploitations agricoles de la France moderne'. *Mélanges offert à Henri Pirenne* (1928).

Idem. The end of the ancient world and the beginning of the Middle Ages (1931).

Idem. La Gaule (1947).

Bibliography

Loyn, H. R. *Anglo-Saxon England and the Norman Conquest* (1962).

Luzzato, G. *An economic history of Italy from the fall of the Roman Empire to the beginning of the 16th century* tr Ph Jones (1961).

McDonald, Donald. *Agricultural writers, 1200–1800* (1908).

McKillop, Alan Dugald. *The background of Thomson's Seasons* (1942).

Madden, D. H. *A chapter of medieval history* (1924).

Mahaffy, J. *A survey of Greek civilisation* (1897).

Maitland, F. W. *Domesday Book and beyond* (1907).

Manitius, M. *Geschichte der Christlich-Lateinischen Poesie bis zur Mitte des 8 Jahrhunderts* (1891).

Idem. *Handschriften antiker Autoren in mittelalterlichen Bibliothekskatalogen* (1935).

Marshall, T. H. 'Jethro Tull and the new husbandry' (*Economic History Review* Jan 1929).

Martial. *Epigraminatum* (III. lviii).

Maxwell of Arkland, Robert. *The practical husbandman* (1757).

Meitzen, A. 'Beobachtungen über Besiedelung, Hausbau und landwirtschaftliche Kultur', *Anleitung zur deutschen und Volks Forschung* (1889).

Michell, H. *The economics of ancient Greece* (1940).

Mills, John. *A new and complete system of practical husbandry* 2 vols (1762).

Mommsen, Theodor. *History of Rome* (1862).

Montaigne. *The complete works* tr Wm Hazlitt. 'Diary of a journey into Italy . . . 1580–1' (1842).

Montelatici, Ubaldo. *Ragionamento sopra i mezzi piu necessari per far refiorire l'agricoltura* (1753).

Montelius, O. *The civilisation of Sweden in heathen times* tr Rev F. H. Woods (1888).

Moretti, K. *Bibliographia agronomica. Saggio d'un catalogo ragionato di libri d'agricoltura e veterinaria scritti in Italiano* (1844).

Moss, H. St L. B. *The birth of the Middle Ages* (1935).

Mousnier, Roland. *Les XVIe et XVIIe siècles. Progress of European civilisation and decline of the Orient, 1472–1715. Histoire générale des civilisations* (IV. 1954).

Münchhausen, Otto von. *Der Hausväter* 3 vols (1765).

Müller, Hans Heinrich. 'Christopher Brown, an English farmer in Brandenberg, Prussia, in the 18th century' (*Agricultural History Review*, 1969, 17, ii, pp 120–135).

Mumford, Lewis. *The human prospect* (1936).

Bibliography

Idem. Tecnics and civilisation (1947).

Musset, L. *Les peuples Scandinaves au moyen âge* (1951).

Musset-Pathay, V. D. de. *Bibliothèque agronomique* (1810).

Neilson, N. *Medieval agrarian economy* (1936).

Nicholson, Edward. *Men and measures: a history of weights* (1912).

Nisard, M. *Les agronomes latins* (1844).

Ogg, David. *Europe of the Ancien Régime, 1715–1783* (1967).

Parain, Ch. 'The evolution of agricultural techniques' (*Cambridge Economic History* I. 1941).

Pardi, del Pero. 'Gli attrezi da taglio per uso agricolo in Italia del antichita fino di giorni nostri in relazione ad una falce con piccolo manico representata in una minatura di un codici manuscritto del XIV secolo existente nel British Museum a Londra' (*Nuovi annali dell' agricoltura* XII. 1. 2. 1933).

Parker, Edward A. & Williams, Ralph M. 'John Dyer, the poet a farmer' (*Agricultural History* XXII. July 1948).

Patullo, M. *Essai sur l'amélioration des terres* (1758).

Paul-Louis. *Ancient Rome at work* (1927).

Perroy, Edouard. *Le moyen âge. L'expansion de l'Orient et la naissance de la civilisation occidentale. Histoire générale des civilisations* (III. 1955).

Piggott, Stuart. *Ancient Europe from the beginnings of agriculture to classical antiquity* (1965).

Pilon, Edmond. *Olivier de Serres* (1924).

Pirenne, Henri. *A history of Europe* (1958).

Idem. Mohammed and Charlemagne tr B. Miall (1958).

Pliny. *Natural history lib. XVII–XIX* tr H. Rackham (Loeb Classical Library, 1961).

Poel, J. M. G. van der. *Heeren en boeren. Een studie over de Commission van Landbouw, 1805–1851* (1949).

Idem. 'Beoefening van de landbouwgeschiedenis in het binnen en buitenland' (*Landbouwgeschiedenis* 1960).

Pollard, Sidney. *The idea of progress. History and society* (1968).

Poni, Carlo. *Gli aratri e l'economia nel Bolognese dell' XVII ad XIX secolo* (1963).

Idem. 'Richerche sugli inventori Bolognese della macchina seminatrice alla fine del secolo XVI' (*Rivista Storica Italiana* LXXVI. 1964).

Poole, A. L. *From Domesday Book to Magna Carta, 1087–1216* (1951).

Bibliography

Pouchet, F. A. *Histoire des sciences naturelles au moyen âge.*

Power, Eileen. 'Peasant life and rural conditions, c 1100–1500' (*Cambridge Medieval History* VII. 1932).

Practical Farmer (David Henry). *The complete English farmer* (1771).

Prato, Giuseppe. *La vita ecomonica in Piemonte a mezzo il secolo XVIII* (1908).

Idem. L'evoluzione agricola nel secolo XVIII (1909).

Prothero, R. E. *The pleasant land of France* (1908).

Prou. M. *La Gaule mérovingienne* (nd).

Pugliese, Salvatore. *Due secolo di vita agricola* (1905).

Putnam, G. H. *Books and their makers during the Middle Ages* 2 vols (1896).

Ramirez, Braulio Anton. *Diccionario de bibliografia agronomica* (1865).

Re, Filippo. *Saggio di bibliografia georgica* (1802).

Idem. Elogio de Piero di Crescenzi detto nella Grand'aula della Reale Università di Bologna (Nov 1811).

Reed, H. S. *A short history of the plant sciences* (1942).

Reed, T. D. *The rise of Wessex* (1947).

Renan, E. *An essay on the age and antiquity of the Book of Nabathean agriculture* (1862).

Renard, G. & Weulersse, G. *Life and work in modern Europe* (1926).

Idem. Le mouvement physiocratic en France (c 1908).

Rhode, Eleanor Sinclair. *The story of the garden* (1933).

Rickard, P. *Britain in medieval French literature, 1100–1500* (1956).

Rivet, A. F. L. *Town and country in Roman Britain* (1958).

Rocque, Auréle la. *The admirable discourses of Bernard Palissy* (1957).

Roger, M. *L'enseignement des lettres classiques d'Ausone à Alcuin* (1905).

Rogers, J. E. T. *Six centuries of work and wages* (ed 1906).

Rostovtzeff, M. *Social and economic history of the Roman Empire* (1926).

Idem. Social and economic history of the Hellenistic world (1941).

Roupnel, G. *Histoire de la campagne française* (1932).

Rück, K. 'Die Naturalis Historia des Plinius im Mittelalter' (*Sitz Bericht D. H. Bayer. Akad d Wiss. Heft 2.* Intro).

Idem. 'Das Excerpt der Naturalis Historia des Plinius von Robert von Cricklade' (*idem* Heft 2. 1902).

Russell, B. *History of western philosophy and its connection with political and social circumstances from the earliest times to the present day* (1947).

Bibliography

Sabbadini, R. *Le scoperte dei Codici Latine e Greci nel' secolo XIV e XV* 2 vols (1905).

Salomon, Noël. *La campagne de la Nouvelle Castille à la fin du XVIe siècle d'après les 'Relaciones topograficas'* (Les Hommes et la terre IX. 1964).

Sambrook, A. J. 'The farmers' boy: Robert Bloomfield, 1766–1823' (*English* no 95, summer 1967).

Sapori, Armando. *Studi di storia economica medievale* (2nd ed 1946).

Savage, Ernest A. *Old English libraries. The making, collection and use of books in the Middle Ages* (Antiquaries Books, 1911).

Savoy, Emile. *Hammurabi à la fin de l'Empire Romain. L'agricultura à travers les âges* (II. 1935).

Sawyer, Charles J. & Darton, F. J. Harvey. *English books, 1475–1900* (1927).

Sayles, G. O. *The medieval foundations of England* (1948).

Schmit, J. C. *Paladii Rutuli Tauri Aemiliani Viri illustris opus agriculturae* (1898).

Schröder-Lembke, G. 'Romische Dreifelderwirtschaft' (*Zeitschrift für Agrargeschichte* XI. April 1963).

Idem. 'Die Hausväterlitteratur als agrargeschichteliche Quelle' (*idem* I. 1953).

Idem. 'Englische Einfluss auf die deutsche Gutswirtschaft in 18 Jahrhundert' (*idem* April 1964).

Idem. 'Die genesis des Colerschen Hausbüches und die Frage seines Quellenwertes' *Wege und Forshungen der Agrargeschichte*. Essays presented to Günter Franz on his sixty-fifth birthday (1965).

Idem. *Zwei frühe deutsche Landwirtschaftschriften: Martin Grosser und Abraham von Thumbschirn* (1965).

Schultz, Theodore W. *Economic crises in world agriculture* (1965).

Schwerz, J. N. *Beschreibung des Landwirtschaft im Nieder-Elsass* (1816).

Seignobos, Charles. *The rise of European civilisation* tr C. A. Philip (1939).

Idem. *History of the French people* (1939).

Sells, A. Lytton. *The paradise of travellers. The Italian influence on Englishmen in the 17th century* (1964).

Semple, Ellen Churchill. *Geography of the Mediterranean region: its relation to ancient history* (1931).

Sidonius, Apollinaris. *Poems and letters* tr W. B. Anderson (Loeb Classical Library 2 vols).

Simon, Joan. *Education in Tudor England* (1966).

Singer, Charles. *A short history of science to the 19th century* (1941).

Sismondi, J. C. L. Sismondi di. *Historical view of the literature of the south of Europe* (2nd ed 1846).

Bibliography

Small, Albion W. *The Cameralists. The pioneers of German social polity* (1907).

Smith, L. M. *The early history of the monastery of Cluny* (1920).

Southern, R. W. *Western views of Islam in the Middle Ages* (1962).

Spencer, T. B. J. 'Lucretius and the scientific poem in English', *Studies in Latin literature and influence* ed D. R. Dudley (1965).

Stancorich, Canon Pietor. *L'aratro-seminatore* (1820).

Steele, R. *Medieval lore from Bartholomew Anglicus* (1905).

Steggall, John H. *A real history of a Suffolk man* (c 1850).

Stenton, F. M. *Anglo-Saxon England* (1950).

Stevens, C. E. 'Agriculture and rural life in the later Roman Empire' (*Cambridge Economic History* I. 1941).

Storr-Best, L. (tr and ed). *Varro on farming* (1912).

Sturgess, R. W. 'The agricultural revolution on the English clays' (*Agricultural History Review* XIV. 2. 1966).

Sutherland, C. H. V. *The Romans in Spain, 217 BC to AD 117* (1939).

Switzer, Stephen. *Ichnographia rustica or the nobleman, gentleman and gardener's recreation* 3 vols (1718).

Taton, R. *A general history of science: ancient and medieval science from prehistory to AD 1450* tr A. J. Pomerans (1963).

Tawney, R. H. & Power, E. E. 'Agrarian life in the Middle Ages' (*Cambridge Economic History* I. 1941).

Taylor, H. O. *The classical heritage of the Middle Ages* (1911).

Idem. The medieval mind 2 vols (1911).

Tenney, Frank. *Economic history of Rome to the end of the Republic* (1920).

Thackeray, W. M. *The English humourists of the 18th century* ed (nd).

Theophrastus. *Enquiry into plants* tr Sir A. F. Gort (Loeb Classical Library).

Thirsk, Joan. 'The common fields' (*Past and Present* no 29. 1964).

Idem. 'Farming techniques', *The agrarian history of England and Wales. IV. 1500–1640* (1967).

Thompson, James Westfall. *The medieval library* (Univ Chicago Studies in Library Science, 1939).

Thompson, J. W. *Economic and social history of the Middle Ages, 300–1300* (2nd ed 1959, 1st 1928).

Thomson, James. *The Seasons* (1726).

Bibliography

Thorndike, Lynn. *Medieval Europe; its development and civilisation* (1920).

Thorpe. *Analecta* (1846).

Tod, Marcus N. 'Economic background of the 5th century BC' (*Cambridge Ancient History* V. 1927).

Toutain, Jules. *Economic life of the ancient world* (1930).

Treneer, Anne. *The mercurial chemist. A life of Sir Humphrey Davy* (1963).

Tull, Jethro. *Horse-hoeing husbandry* (4th ed 1762).

Usher, A. P. *A history of mechanical inventions* (1954).

Vaschelde, Henri. *Olivier de Serres, Seigneur de Pradel. Sa vie et ses travaux* (1866).

Villard, Eugène. *Olivier de Serres et son oeuvre* (1872).

Vinogradoff, P. *Growth of the manor* (1905).

Idem. English society in the 11th century (1908).

Wallace-Hadrill, J. M. *The long haired kings and other studies in Frankish history* (1962).

Waltz, Pierre. *Hesiod et son poème moral* (1906).

Way, Ruth & Simmons, Margaret. *A geography of Spain and Portugal* (1962).

White, Kenneth D. 'The efficiency of Roman farming under the Empire' (*Agricultural History* XXX. April 1956).

White Jr, Lynn. *Medieval technology and social change* (1962).

White, R. J. *Europe in the 18th century* (1965).

Whitelock, Dorothy. *The beginnings of English society* (Pelican History of England. II. 1954).

Wilkinson, L. P. 'The intention of Virgil's Georgics' (*Greece and Rome* XVIII. 55. Jan 1950).

Wilson, A. Stephen. *A bushel of wheat* (1883).

Winter, George. *A new and comprehensive system of husbandry* (1787).

Wood, P. D. 'Strip lynchets reconsidered' (*Geographical Journal* 127. 4. 1961).

Woolf, S. J. 'Economic problems of the nobility in the early modern period' *Economic History Review* 2nd ser. XVII. 2. Dec 1964).

Wright, Thomas. *The Celt, the Roman and the Saxon* (1852).

Idem (ed). *A volume of vocabularies from 10th to 15th centuries* (1857).

Xenophon. *Treatise of householde* tr Gentian Hervet (1499–1584). *Certain antient tracts concerning the management of landed property reprinted* (1767).

Young, Arthur. *Travels in France* ed Miss Betham-Edwards (ed of 1905).

II

Works consulted, but not specifically mentioned in the Notes, and therefore not in the Bibliography.

Ackroyd, W. R. *Three philosophers* (1935).

Berthelet, P. E. M. *La révolution chimique Lavoisier* (1890).

Borheck, Georg Heinrich. *Entwurf einer Anweisung zur Landbaukunst* (1792).

Idem. Lehrbuch der Landbaukunst für Baumeister und Landwirt 2 vols (1822).

Richard, Bradley. *New improvements of gardening and planting* (3rd ed corrected 1719).

Idem. Ten practical discourses concerning earth, water, fire and air as they relate to the growth of plants (1729).

Bruford, W. H. *Germany in the 18th century* (1965).

Buckle, H. T. *History of civilisation in England* 3 vols (ed of 1873).

Calonne, le Baron A. de. *La vie agricole sous l'ancien régime en Picardie et en Artois* (1883).

Carter, H. B. *His Majesty's Spanish flock* (1964).

Clark, Roy. *Black sailed traders; the keels and wherries of Norfolk and Suffolk* (1961).

Clarke, Desmond. *The unfortunate husbandman* Charles Varley (1964).

Clark-Kennedy, A. E. *Stephen Hales, DD, FRS. An 18th century biography* (1929).

Colman, Gould P. 'Innovation and diffusion in agriculture' (*Agricultural History* XLII. 3. July 1968).

Crofts, J. 'Wordsworth in the 17th century' (Warton lecture in English poetry, 21 Feb 1940. *Proc British Academy* XXVI).

Crosland, Maurice P. *Historical studies in the language of chemistry* (1962).

d'Aubenton. *Instruction pour les bergères et pour les proprietaires des troupeaux* (3rd ed. An X. 1st 1782).

Davis, Stella. 'The agricultural history of Cheshire' (*Chetham Society* 3rd ser. X. 1960).

Goldsmith, Oliver. *An history of the earth and animated nature* 1774. Selected works chosen by Richard Garnett (1950).

Harrison, Gustavus. *Agriculture delineated, or the farmers' complete guide* (1775).

Hart, A. Tindall. *Country counting houses. The story of two 18th century clerical account books* (1962).

Bibliography

Hirschfeld, G. G. L. *Das Landleben* (3rd ed 1771).

Home, Henry (Lord Kames). *The gentleman farmer* (1776).

Johnson, A. H. *The disappearance of the small landowner*, intro by Joan Thirsk (1963, 1st 1909).

Jones, E. L. & Mingay, G. E. *Land, labour and population in the Industrial Revolution. Essays presented to J. D. Chambers* (1967).

Kent, Nathaniel. *Hints to gentlemen of landed property* (2nd ed 1776).

Kerr, Barbara. *Bound to the soil* (1968).

Lisle, Edward. *Observations on husbandry* (1757).

Masefield, John. *The country scene* (1937).

Mathias, Peter. *The brewing industry in England, 1700–1830* (1959).

Meuvret, J. 'L'agriculture en Europe au XVIIe et XVIIIe siècles', *Relazione X Congresso Internationale di Scienza Storiche* (IV. 1955).

Mingay, G. E. *English landed society in the 18th century* (1963).

Parkinson, Richard. *The experienced farmer enlarged and improved* 2 vols. (2nd ed. 1807).

Poel, J. M. G. van der. *Wegwijzer in de landbouwgeschiedenis* (1953).

Raistrick, Arthur. *Quakers in science and industry* (1950).

Read, John. *Prelude to chemistry. An outline of alchemy, its literature and relationships* (1936).

Riemann, Friedrich Karl. 'Ackerbau und Viehhaltung im vorindustrielle Deutschland', *Beiheft zum Jahrbuch der Albertus Universität zu Königsberg* (III. 1953).

Robertson, J. G. *The life and work of Goethe, 1749–1832* (1932).

Sickler, Johann Volkmar. *Die deutsche Landwirtschaft in ihren ganzen Umfangen nach der neuesten Erfahrungen* 4 vols (1803).

Smeaton, W. A. *Fourcroy, chemist and revolutionary, 1755–1809* (1960).

Smith, J. H. *The Gordon's Mill farming club, 1758–64* (1962).

Sneller, Z. W. *Geschiedenis van de Nederlandse landbouw, 1795–1940* (1951).

Spahr, J. J. van der Hoek. *Geschiedenis van de Friese landbouw* (1952).

Stockdale, James. *Annals of Cartmel* (1872).

Vermale, François. *Les classes rurales en Savoie au XVIIIe siècle* (1911).

Wilson, Charles. *Holland and Britain* (c 1960).

Wolters, Fritz. 'Studen über die Agrarzustande und Agrarprobleme in Frankreich von 1760 bis 1790', (*Schmollers Stadts-und Sozialwissenschaftsliches Forschungen* 21. pt 5 1905).

Bibliography

III

Works mentioned in the text but not included in either of the above lists.

Addison, Joseph. *Spectator* (1711).

Aelfric. *Colloquies* (tenth century).

Africo, Clemente. *Dell'agricoltura accomodate all'uso die nostri tempi* Lib VI. (1572, etc).

Agricola, Georg Andreas. *Neu und nie erhörter doch in die Natur, 1716–17* (tr English by Bradley 1720, Dutch 1724).

Agricola, J. Jac. *Schauplatz* (1676 and 1678).

Alamanni, Luigi. *La coltivazione* (1546).

Albertus Magnus. (1206–80) *de Vegetabilibus*.

Albin, Moller. *Der grosse alte Schreitkalendar* (1605).

Alstrom, Col. *Essai sur la race des brebis á laine fine* (1773 tr from theoriginal Swedish).

Ambrogini, Angel (Politian). *Sylva*, lectures on Hesiod and Virgil (sixteenth century).

Anon. *The crafte of graffynge and plantynge of trees* (c 1500).

Idem. *Dos discorsos solve el Gobierno de los granos y cultivo de las tierras* (1775).

Idem. *L Junius Moderatus Columella of husbandry in twelve books* (tr into English 1745).

Idem. *Quatre traitez utilis et delectabilis de l'agriculture* (1560).

Idem. *Pflanzbuchlein* (sixteenth century).

Idem. *Agricultura practica* (1626).

Idem. *Ricordi di agricoltura raccolti da migliori autori di coltivazione antichi e moderni* (3rd ed 1735).

Idem. *Rustici latini volgarizzati* (1792–1800).

Idem. *Virgil's Lehrbuch von der Landwirtschaft* (1792). Also tr into German by Johann Heinrich Jacobi (1797).

Anton, Karl Gottlob. *Geschichte der teutschen Landwirtschaft* (1799).

Aquinas, Carolus. *Nomenclator agriculturae* (1736).

Arcere, M. *De l'état de l'agriculture chez les Romains* (1777).

Aristotle. tr as *Historia general de aves y animales* by Diego de Funes y Mendoza (1621).

Banqueri, Josef Antonio. *Libro de agricultura* (1802) tr of ibn al Awam.

Bibliography

Barthes, M. *Mémoire d'agriculture et de la mécanique* (1763).

Barpo, Gio Battista. *Le delitie e i frutti dell'agricoltura e della villa, libri tre* (1634).

Bartholomew, Anglicus. *De proprietatibus rerum* (c 1230–40).

Bellot, James. *The Booke of thrift conteyning a perfite order and right maner to profite lands* (1589) (really Walter of Henley).

Belon, Pierre. *Les remonstrances sur le défaut du labour et culture des plantes* (1558).

Benini, Vincenzo. *Notes on Alamanni with the twelve books of Crescentius added* (1745).

Berch, Andreas. *Methodus investigandi origines gentium ope instrumentum rurali* (1795).

Bergerie, Rougier de la. *Histoire de l'agriculture française* (1815).

Bertrand, J. *Elémens d'agriculture fondés sur les faits et les raissonnements à l'usage du peuple de la campagne* (1775).

Biondino, Mich Ang. *Historia della pianti de Theophrasto libri III* (1548).

Boeckler. *Haus- und Feldschule* (1686, etc).

Bonardo, Gio Marcia. *Le richezza dell'agricoltura* (1584, etc).

Bonneterie, Saboureux de la. *Economie rurale par Caton, Varron, Columelle Palladius et Vegece* (1772).

Bradley, Richard. *Survey of the ancient husbandry and gardening* (1725).

Idem. The science of good husbandry or the Oeconomics of Xenophon (1727).

Brie, Jehan de. *Le bon berger* (MS 1379, printed c 1540).

Cacherano, G. F. M. *De mezzi per introdurre . . . la coltivazione e la populazione nell'agro Romano* (1785).

Camerarius, Joachim. *De re rustica, opuscula nonnulla, lectu cum jucunda tum utilia, jam partim premium composita, partim edita* (1577).

Caraccioli. *L'agriculture simplifiée d'après les règles des anciens* (1749).

Chomel, Noel. *Dictionnaire Oeconomique* (1709. tr English by Richard Bradley 1725; into Dutch 1748 and 1778–93).

Choyselat, Prudent. *Discours Oeconomique* (1569).

Clarke, Cuthbert. *True theory and practice of agriculture* (1777. tr French 1779).

Coler, Johann. *Kalendar* (1591).

Idem. Oeconomica ruralis et domestica (1593, 1601).

Columella. tr as *Les douze livres . . . des choses rustiques* (1552, etc).

Corradi, Bernadino. *Versione Italiana del decimo libro di L. G. Moderato Columella* (1754).

Costa, M. le Marqués. *Essai sur l'amélioration de l'agriculture dans les pays monteux . . . Savoie, Chambery . . .* (1774).

Bibliography

Curtius, M. C. tr *Columella* (in German) (1769).

Desplaces, L. B. *Preservatif contre l'agronomie* (1762).

Idem. Histoire de l'agriculture ancienne extrait de l'Histoire Naturelle de Pline, Livre XVIII (1765).

Dickson, Adam. *Treatise on agriculture* 2 vols (1762, 1770).

Idem. Husbandry of the Ancients 2 vols (1788, tr French 1801-2).

Diderot. *Lettre écrite de la Comté du Norfolk.*

Dumas, G. (tr). *Economique de Xenophon et de projet de finance du même auteur* (1768).

Eckhart, Johann Gottlieb. *Vollständige Experimental-ökonomie über das vegetabil-animalische* (1754).

Engelen, C van. *De nieuwe Wijze van Landbouwen* 4 vols (1762-5). Mainly a tr of du Hamel.

Estienne, Charles. *Praedium rusticum* (1554).

Idem and Liebault, Jean. *L'agriculture et maison rustique* (1567, etc). tr as: *L'agricoltura e casa da villa* (1581). *De veltbauw ofte landwinninghe* (1588). *Maison rustique or the countrey ferme* (1600 by Richard Surflet). *The country ferme* (1616 by Gervase Markham).

Evelyn, John. *Sylva* (1664). *Terra* (1676).

Fabroni. *Istruzione elementari de agricoltura* (1786).

Fellemburg, Emanuel. *Vues relatif à l'agriculture Suisse et aux moyens de la perfectionner* (1808).

Fischer. *Haus, etc* (1696).

Fitzherbert. *Book of husbandry* (1523).

Florini, Francisci Phillippi (Pfalzgraf Franz Phillip bei Rhein). *Oeconomus prudens et legalis* (1751).

Fraas, Carl. *Geschichte der Landwirtschaft, 1750-1840* (1852).

Fresne, Ebaudy de. *Traité d'agriculture* 2 pts (1788).

Gallo, Agostino. *La dieci giornata dell vera agricoltura* (1556). tr as *Secrets de la vraye agriculture* (1571).

Idem. La vinti giornata dell'agricoltura (1573). There were other versions of this treatise as ten and other numbers of days.

Gassers, Simon Peter. *Einleitung zu den Oeconomische, Politische, und Cameralwissenschaften . . .* (1729).

Gesner, John Mathias (ed). *Scriptores rei rusticae* 2 vols (1735).

Ginanni, Francesco. *Della malattie del grano in erba. Tratto storicafisica* (1759).

Ginzrot. *Die Wägen und Fuhrwerke der Greichen und Römer* (1817).

Bibliography

Goëlle, Dammertin de. *Le plaisir des champs* (sixteenth century).

Grosser, Martin. *Kurze und gar einfeltige Anleitung zu der Landwirtschaft* (1590).

Guyon, Loys 'Sieur de la Mauche. *Les divers leçons* (1604).

Gyllenborg, Count Gustavus Adolphus. *Agriculturae fundamenta chemica* (1761). tr German 1764; French 1766; English 1770; Latin in France 1791; Spanish 1794.

Hagedorn, Hofrat. *Landwirtschaftlicher Haushalter* (1755).

du Hamel, du Monceau. *Traité de la culture des terres* 6 vols (1750-6).

Hartlib, Samuel. *His Legacie* (1651, etc).

Heresbach, Conrad. *Rei rusticae libri quattuor, universam rusticam* (1570). tr English, Barnaby Googe (1577, etc); Gervase Markham (1631); reptd as *The perfect husbandman* (1658).

Hermann, Chr. *Haushaltungsbuch* (1674, 1677).

Herrara, Gabriel Alonzo de. *Libro de agriculture* (1539, etc).

Home, Francis. *Principles of agriculture and vegetation* (1757). tr French 1761; German 1763.

Kastos, Hadji Khalfa. *Kitab al felihah ar roumieh*, Book of Greek agriculture (? tenth century).

Kembter, Adriano. *Veterum scriptorum de re rustica praecepta in dialogus collecta* (eighteenth century).

Kulbel, Dr M. *Dissertation sur la cause de la fertilité des terres* (1741).

Laporta, Francisco Luis. *Historia de la agricultura España* (1798).

Lasteyrie, Graf von. *Sammlung von Maschinen, Instrumenten und Geräteschaften, etc* (1821-3).

Leopoldt, Johann Georg. *Nützliche und auf die Erfahrung gegründete Einleitung zu der Landwirtschaft* (1759).

Liger, Louis. *La nouvelle maison rustique* (1702, etc, tr Spanish 1720).

Maggazini, P. D. Vitale. *Coltivazione Toscana . . .* (1634).

Maimonides, Moses. (1135-1204) *More novochim*.

Marshall, William. tr French as *Maison rustique anglais ou voyage agronomique en Angleterre* 5 vols (1806).

Martyn, John, FRS, tr *Virgil Georgics* (1740-1, etc).

Mascall, Leonard. *Government of cattel* (1596, etc), enlarged by Richard Ruscam (1680).

Mason, George. *An essay on design in gardening . . .* (1768) cited by Wordsworth as *English Garden*.

Milizia, Brazuolo Paolo. *Esiodo. Le Opere e i Giorni tradotto* (1765).

Bibliography

Milton, John. *Tractate on education* (1642).

Moller, Tobias. *Sommer Feldbau* (1583). *Winter Feldbau* (1584).

Mongez. *Mémoire sur les instruments d'agriculture des anciens* (1815).

Mortimer, John. *Whole art of husbandry* 2 vols (1707. tr French 1765).

Needham, Peter. *Geoponika*, Greek and Latin text (1704).

Nenci, Giuseppe. *Reflessi sopra le piu frequenti e necessari operazione della coltivazione* (1691, 1796).

Paoletti, Ferdinando. *Pensieri sopra l'agricoltura* 2 vols (2nd ed 1789).

Philipps, John. *Cyder* (1706. tr into French and Italian).

Plat, Sir Hugh. *The jewell house of art and nature* (1594).

Idem. Sundrie new and artificial remedies against famine (1596).

Pluche, de la. *Spectacle de la nature* 8 vols (1754-5).

Pope, Alexander. *Discourse on pastoral poetry*.

Quiqueron, Pierre de, Bishop of Senes. *La nouvelle agriculture* (tr by Francis de Claret 1613, etc).

Rau, K. H. *Geschichte des Pfluges* (1845).

Reichardt, Christian. *Land-und Gartenschatz* 6 vols (1753-5).

Rosa, Gabriella. *Storia dell'agricoltura nell civilla* (1883).

Rosco, M. *Agricoltura tratto di diversi antichi e moderna scrittori di lingua Spagnuola 1568.* tr of Herrara.

Rosenow, C. F. *Versuch einer Abhandlung von Ackerbau und Koppelwirtschaft* (1762).

Rozier, Abbé. *Nouveau cours complet d'agriculture* 6 vols (1809).

Rucellai, S. Giovanni. *L'api* (1590).

Rueneuve, Angran de. *Observations sur l'agriculture et le jardinage* 2 vols (1712).

Sabino, Troilo. *Praelictus in Virgilie Georgica* (1526).

Schneider, Gottlop (ed). *Scriptorum rei rusticae veterum Latinorum* (1794-7).

Schronius, Wolfgang Adolf. *Syntagma de rebus rusticus et oeconomicus ex rei rusticae scriptores conscriptum* (1735).

Schultze, Fr G. *De aratri Romain forma et compositione* (1820).

Schumacher, C. W. *Gerechte Verhältnisse der Viehzucht zum Ackerbau aus de verbesserten Mecklenburgische Wirtschaftsverfassung* (1763).

Scott, Sir Walter. *The Pirate*.

Seabirgiro, Giulio Cesare. tr *Aristotle de plantis* (1566), *Theophrastus* (1584).

Serres, Olivier de. *Le théâtre d'agriculture et mesnage des champs* (1600).

231

Bibliography

de Sotomayer y Rubio. *Doce libros de agricultura que escribo in Latin . . . Columella* (1824).

Stella, Benedetto. *Il tobacco* (1669).

Tanara, Vincenzo. *L'economica del citadino in villa* (1651).

Tarello, Camillo. *Ricordo d'agricoltura* (1567, etc).

Thaer, Albrecht. *Einleitung zur Kenntnis der englischen Landwirtschaft* 3 vols (1804).

Thieme, Joh Chr. *Haus-Feld-Arzney, Koch, Kunst und Wunder-buch* (1682).

Thumbschirn, Abraham von. *Oeconomica oder notwendiger Unterricht und Anleitung wie eine ganze Haushaltung* (1616. New edn ed Gertrud Schröder-Lembke 1965).

Thunen, Johann Heinrich von. *Der isolierte Staat* (1826, etc).

Toaldo, D. Giuseppe. *La meteorologica applicata all'agricoltura* (new ed 1786).

Trinci, Cosimo. *L'agricoltura sperimentato ovvero regolo generali sopra l'agricoltura* (1726, reptd 1851).

Idem. Raccolta d'opusculi appartenendi all'agricoltura (1768).

Idem. Nuova trattato d'agricoltura (1778).

Turbilly, Marquis de. *Mémoire sur les defrichements* (1760).

Vallemont, Abbé de. *Curiositez de la nature et de l'art sur la vegetation* (1703. tr by Bishop William Fleetwood into English 1707).

Valligo, José Manuel Fernandez. *Prados artificiales* (1797).

Vanieri, Jacobi. *Praedium rusticum* (1696).

Varlo, Charles. *New system of husbandry* 3 vols (1770. tr French 4th ed 1775).

Venuto, Antonino. *L'agricoltura . . . Campi, Prata, Orti, Giardini, Viti, Arbori* (1516).

Idem. De agricoltura opusculum (1537).

Vinet, Elie and Molluson, Antoine Mizauld de. *La maison champestre et agriculture* 5 parts (1602).

Weston, Sir Richard. *A discours of husbandry used in Brabant and Flanders* (?1650).

Wolfius, Dr. *A discovery of the true cause of the wonderful multiplication of corn* (1734).

Worlidge, John. *Systema agriculturae* (1669).

Young, Arthur. *Political arithmetic* (tr into French 1774).

Idem. Travels in France (1789 et seq. tr into French).

Zincke, Georg Heinrich. *Allgemeines ökonomisches Lexikon* (1744).

Index

References to books become so numerous after the fifteenth century that these have been grouped under the relevant country and period, eg, England, text-books, sixteenth century. Specific authors and titles can be found in the Bibliographies (p 210-32). References to crops appear as, eg, wheat, *see* Crops, cereal.

Index